T0262145

Micro-Nano Electrochemical Systems and Fabrication Techniques Handbook

Micro-Nano Electrochemical Systems and Fabrication Techniques Handbook

Edited by **Eve Versuh**

New York

Published by NY Research Press,
23 West, 55th Street, Suite 816,
New York, NY 10019, USA
www.nyresearchpress.com

**Micro-Nano Electrochemical Systems and
Fabrication Techniques Handbook**
Edited by Eve Versuh

© 2015 NY Research Press

International Standard Book Number: 978-1-63238-327-3 (Hardback)

This book contains information obtained from authentic and highly regarded sources. Copyright for all individual chapters remain with the respective authors as indicated. A wide variety of references are listed. Permission and sources are indicated; for detailed attributions, please refer to the permissions page. Reasonable efforts have been made to publish reliable data and information, but the authors, editors and publisher cannot assume any responsibility for the validity of all materials or the consequences of their use.

The publisher's policy is to use permanent paper from mills that operate a sustainable forestry policy. Furthermore, the publisher ensures that the text paper and cover boards used have met acceptable environmental accreditation standards.

Trademark Notice: Registered trademark of products or corporate names are used only for explanation and identification without intent to infringe.

Printed in the United States of America.

Contents

Preface

Over the recent decade, advancements and applications have progressed exponentially. This has led to the increased interest in this field and projects are being conducted to enhance knowledge. The main objective of this book is to present some of the critical challenges and provide insights into possible solutions. This book will answer the varied questions that arise in the field and also provide an increased scope for furthering studies.

This book is a collection of researches by various experts who share their analysis of current trends and future implications of micro-nano electrochemical systems. MEMS technology is progressively influencing and improving our quality of life. Simultaneously, progress in nanotechnology and nanomaterials has accelerated the growth of NEMS. This book consists of an analysis of modern technologies and their forthcoming developments. It focuses on the recent progress in devices and production procedures in the field of these miniaturized electromechanical structures. This book presents information about design, production and packaging; as well as solutions in these aspects for targeted functions. This book aims to help experts and students who are engaged in this field.

I hope that this book, with its visionary approach, will be a valuable addition and will promote interest among readers. Each of the authors has provided their extraordinary competence in their specific fields by providing different perspectives as they come from diverse nations and regions. I thank them for their contributions.

Editor

Liquid Encapsulation Technology for Microelectromechanical Systems

Norihisa Miki

Additional information is available at the end of the chapter

1. Introduction

Microelectromechanical systems (MEMS) have been extensively studied for over three decades, which has resulted in the prevalence of quite a few commercially available MEMS products in our daily lives, although they are too small to see. In the very beginning of the MEMS success story, people recognized the importance of packaging [1]. MEMS contain mechanical parts, and given their small sizes, they are severely affected by surrounding molecules. Therefore, MEMS are packaged under vacuum, at low pressure, or at least free from water molecules. Water molecules can bridge two separated parts and bring them into contact by the meniscus force, which may lead to permanent adhesion of the parts, known as stiction. This phenomenon must be averted, not only in the packaging, but also in the fabrication of parts. It is not an overstatement to say that researchers go to great lengths to keep their devices dry.

On the other hand, as MEMS technologies advance, a wide variety of applications are expected, some of which the MEMS must handle liquids. For example, drug delivery systems (DDS) that administer medicine to diseased parts at designated times can employ MEMS that are sufficiently small to be implanted and are capable of controlling discharge of the medicine [2-5]. In this application, the MEMS must contain medicine, which is in liquid form in many cases. In addition, MEMS can be used as a portable power source, referred to as power MEMS. Micro gas turbines and certain fuel cells require liquid fuel to generate chemical reactions [6-9]. Micro total analysis systems, or microTAS, are used to manipulate minute aqueous analytes and/or control microfluids to handle samples, such as cells and bacteria, for biochemical analysis [10-14].

Such useful characteristics of liquids are available to expand the design space for innovative MEMS devices. Functional liquids, such as magnetorheological fluid and electroconjugate liquids, can be used as micro pumps and actuators [15-16]. Other useful characteristics of liquids include deformability, incompressibility, and high dielectric constant. Hydraulic amplification can be achieved by exploiting the deformability and incompressibility of liquids [17-22]. In addition, highly dielectric liquids can enhance sensor sensitivity while maintaining flexibility [23].

While some applications allow such MEMS devices to bring the liquid from outside, encapsulation of liquid inside MEMS devices is mandatory in other applications. Liquid encapsulation technology can be used to manufacture innovative MEMS devices, such as completely spherical microlenses and hydraulic amplification mechanisms. Various liquid encapsulation technologies have been proposed to achieve these promising applications. The liquid species to be encapsulated and the application must be taken into consideration for the selection of appropriate encapsulation processes. This chapter reviews state-of-the-art liquid encapsulation technologies and their application to the manufacture of innovative MEMS devices that exploit the useful characteristics of the encapsulated liquids.

2. Fill and seal technique

The most straight-forward technique to encapsulate liquids is to dispense liquid into a reservoir and then seal it with another substrate, as shown in Figure 1. The reservoirs can be easily manufactured using conventional MEMS fabrication technologies. Currently, commercially available dispensers are capable of dispensing a minute amount of liquid, as small as several nanoliters. Sealing, or bonding the substrates, is the most critical process.

In the MEMS field, bonding technologies have been widely explored for packaging and manufacturing three-dimensional structures [24-27]. The direct bonding of silicon wafers, anodic bonding of glass substrates, and thermocompression bonding using a metal thin film as an adhesive layer are examples of frequently used technologies. These technologies achieve strong bonding of substrates via covalent bonds; however, these processes have drawbacks when applied to the sealing of a liquid filled reservoirs if they require high temperatures. High temperature processes in the order of several hundreds degrees Celsius may change the properties of the liquid to be encapsulated. For example, in drug delivery systems, it is necessary to maintain the medicinal effect of the encapsulated drug. Fuels for power MEMS applications should not be burnt before the device is completed. In addition, some liquids are volatile, which precludes not only high-temperature, but also vacuum processes.

Therefore, in this fill and seal approach, adhesive bonding shown in Figure 1 is the most appropriate technique. The adhesives employed include epoxy, UV curable resin including photoresist, and benzocyclobutene (BCB). Such adhesives solidify after either mixing with curing agents, exposure to UV irradiation, or thermal treatment at low temperatures. Either one-part or two-part epoxy resins can be used, because they do not require high temperature to promote solidification. The chemical reaction progresses with time, even at room temperature, and after solidification, the epoxy achieves strong bonds and is resistant to many chemicals.

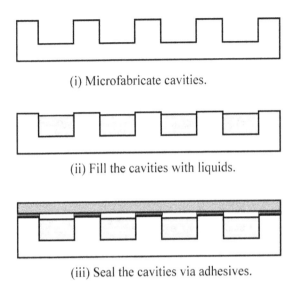

(i) Microfabricate cavities.

(ii) Fill the cavities with liquids.

(iii) Seal the cavities via adhesives.

Figure 1. Fill and seal approach.

While epoxy is a common adhesive, it is not compatible with conventional MEMS fabrication technology. To achieve reliable and reproducible bonding, the adhesives are preferably spin-coated, which allows the thickness to be controlled according to the spinning speed. In this regard, photoresist is a good candidate. Photoresists are compatible with MEMS fabrication technologies and can be spin-coated, and more importantly, knowledge of their use is well developed. Photoresist is coated on a substrate and then brought into contact with the pairing substrate either before or after the curing processes. A typical curing temperature is around 100 °C. When the contact is performed after curing, the bonding is achieved by a hot melt process at higher temperatures, although lower than 200 °C. Photoresist can be patterned using conventional photolithography to determine the bonding areas. The major drawback of using photoresists as adhesives is the weakness of the bond strength, i.e., they are not designed to function as adhesives. Adhesion between the photoresist and the substrates, as well as the mechanical strength of the photoresist, is designed to be sufficiently strong to survive photolithography processes. Therefore, the bonding may fail due to external forces, either at the interface or within the bulk.

BCB is a promising polymer adhesive that is photo-patternable and compatible with conventional photolithography processes. It can be spin-coated to thicknesses of 5-15 μm at spinning speeds of 1000-6000 rpm [28]. BCB has good chemical resistance, and the most significant advantage of this material is that it does not release any gases during the cure, which does not create pores in the material or contaminate the encapsulated liquid. BCB can be used to bond two substrates by thermocompression bonding. Compression at 230 °C has been attempted, which may limit the species of liquid to be encapsulated. However, BCB has been applied to seal sodium hypochlorite aqueous solution (NaOCl) for galvanic cells [28]. The paper discusses the BCB thickness and the bond quality determined by the geometry of the bonding areas.

UV curable resins do not require heat treatment, but only UV irradiation. If the MEMS devices are not UV sensitive and one substrate is transparent to UV light, then UV curable resin offers a strong bond after solidification with UV irradiation. Such bonding can even be conducted in liquids [21,22,29] and we have termed this the bonding-in-liquid technique (BiLT).

We have introduced sealing processes that employ polymer adhesives. However, the gas permeable nature of polymers may cause problems of contamination and vaporization of volatile liquids. For example, polydimethyl siloxane (PDMS), which is one of the most frequently used polymers in the fields of MEMS and microTAS, is permeable to gas. However, this permeability can be modified by the addition of different materials [30] or coating with airtight films [31]. Typical polymers are several orders of magnitude more permeable to gas than metals and ceramics [27]. Therefore, sealing with gold stud bumps has been proposed [32], where reservoirs are filled up with the liquids via microchannels and the inlets and outlets of the channels are then plugged with wire-bonding gold. Firstly, a gold ball is formed at the edge of the gold wire by electrical discharge. The ball is then pressed to the opening of the channel using ultrasound. The wire is then cut and the sealing is completed. Helium leak tests were conducted and hermetic sealing was verified using this technique when the hole diameters were less than 40 μm.

The inevitable drawback of the fill and seal approach is the filling rate; it is quite difficult to completely fill a reservoir with a liquid. This is acceptable for some applications, such as drug delivery and fuel supply for power MEMS devices. However, the performance of hydraulic displacement amplification mechanisms (HDAM) is deteriorated by the interfusion of compressible air. When liquids are used as components of sensors, contamination of gas or other liquids will lead to a loss of sensitivity. Therefore, liquid encapsulation techniques that enable complete filling of liquids are mandatory. The author's group developed BiLT, which is a fill and seal approach that enable complete filling [21,22,29].

3. Bonding-in-Liquid Technique (BiLT)

HDAMs require complete filling of the reservoir with an incompressible liquid, because gas is much more compressible than liquid. Figure 2 shows a package chamber that has openings at the top and bottom, where incompressible liquid is encapsulated with flexible polymer membranes. The top opening, which is determined by a metal plate, is smaller than the bottom opening; therefore, a small displacement applied to the bottom membrane is amplified at the top, according to the ratio of the openings. The application of HDAM is discussed in section 5.2. The key points in the fabrication of HDAMs are no interfusion of air bubbles and sealing with flexible polymer membranes. Complete filling can be achieved by the direct deposition of a thin film, which is detailed in the following section; however, this technique does not allow the use of flexible membranes. We have developed BiLT [29], which can be employed to overcome this problem.

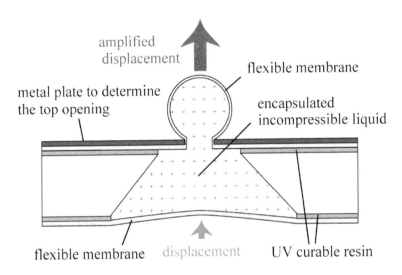

Figure 2. Schematic cross-sectional view of HDAM.

Rather than package the MEMS devices vacuum, we considered that if the encapsulating process was conducted in liquid, then the reservoir could be filled without the interfusion of air bubbles. However, one concern was how to successfully bond the membrane to the package chamber in a liquid environment. Therefore, it was decided to use a UV-curable resin (3164 Three Bond, Three Bond Co., Ltd.) that is solidified after UV irradiation, even in a liquid environment. This membrane achieves a tensile strength of 0.85 MPa when cured and the thickness of the resin can be controlled according to the spin-coating speed.

Figure 3 depicts the procedures employed in BiLT. Firstly, a UV resin is coated onto the bonding surface; however, it should be noted that the surface has many cavities for liquid encapsulation. Therefore, the UV resin is spin-coated onto a thick PDMS membrane and then transferred onto the bonding surface by soft contacting the PDMS membrane (Figure 3(a,d)). A sufficient amount of resin needs to be applied to the bonding surface to achieve a good bond, while excess resin may fall into and occupy the cavity during the bonding process. UV resin thicknesses of 80, 120, 160, and 200 μm were tested on PDMS membranes, which correspond to spin-coating speeds of 4000, 3000, 2000, and 1000 rpm, respectively. When silicon was used as the bonding substrate, the transferred thicknesses were 7.9, 8.1, 17, and 27 μm. In case of UV resin thicknesses of 17 and 27 μm, the excess resin flowed into the cavity.

Handling of a flexible thin membrane is not a trivial process. The membrane must be kept flat throughout the bonding process. Therefore, the membrane was spin-coated and cured on a glass substrate. The thickness of the PDMS membrane can be controlled according to the spin-coating speed. During the bonding process, the PDMS membrane must be peeled off the glass substrate. Therefore, the glass surface is coated in advance with a hydrophobic film (CYTOP M, H, Asahi Glass Corporation) to facilitate exfoliation.

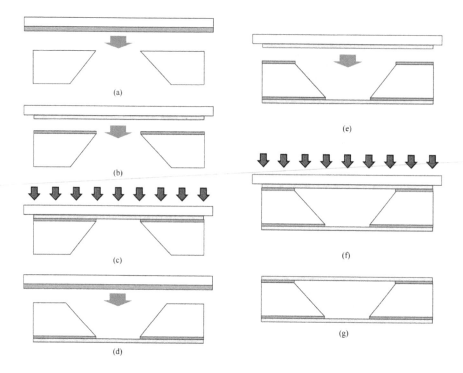

Figure 3. BiLT process. (a) UV curable resin is transferred onto the bonding surface. (b) A flexible membrane coated onto another substrate in advance is brought into contact with the bonding surface. The membrane is coated onto a hydrophobic layer to facilitate peeling of the membrane from the substrate. (c) UV light is irradiated to cure the UV resin. (d) A second UV curable resin is transferred onto the bonding surface. (e) A flexible membrane is brought into contact with the bonding surface in a liquid environment. (f) UV light is irradiated to cure the resin. (g) Liquid encapsulation without the interfusion of air bubbles or deformation of the membrane is achieved.

The substrate with cavities and the flexible membrane on the glass substrate are brought into contact in a liquid environment (Figure 3(e)). UV light is then irradiated onto the bonding surface through the glass substrate and flexible membrane to cure the UV curable resin (Figure 3(f)). Note that the substrate and membrane must be UV-transparent for this process. Figure 4 shows micrographs that confirm liquid (red-dyed deionized (DI) water) encapsulation was completed without the interfusion of air bubbles. Excess resin flowed into the cavities for UV resin thicknesses of 17 and 27 μm. No DI water was observed at the bonding interface. Encapsulation of glycerin was also attempted. Glycerin is non-volatile, so that the volume of the encapsulated glycerin did not change over a period of weeks even when encapsulated with a gas permeable PDMS membrane at ambient pressure and room temperature.

In HDAM, encapsulated liquids are sealed with flexible membranes at both the top and bottom sides of the package chamber. When the encapsulation/bonding process is conducted in air and not in liquid, the difference in the density of the air and liquid result in bowing of the membrane. Note that the membranes must be kept flat during the bonding processes of BiLT.

The bond strengths were investigated by conducting 180° peel tests on PDMS membranes and silicon substrates bonded using BiLT in DI water, glycerin, phosphate buffer solution (PBS), isopropyl alcohol (IPA), and acetone, and also in air as a reference. The silicon substrate used in the experiments did not contain bonding cavities. The bond strengths of the samples were measured as a function of time (1, 6, 24, 72, and 168 h) using a dynamic mechanical analyzer (RSAIII, TA Instruments). The test procedure involved one edge of the PDMS membrane being manually peeled from the silicon substrate and the unbonded area of the silicon substrate being clamped. The peeled PDMS membrane was then pulled in the direction parallel to the bonding interface at a speed of 3 mm/min until it peeled off, and the shear stress required to peel the PDMS membrane from the silicon substrate was measured. The results are shown in Figure 5. The bonding resin was dissolved in both IPA and acetone solution, and thus bonding was unsuccessful when conducted in these solutions, while the bonding strengths of the other samples were comparable. The bond strengths increased with time, most likely due to continuing chemical reaction of the UV-curable resin over time. The bonding strengths after 1 week were more than 4 times greater than those obtained after 6 h. Peel tests conducted within 72 h of bonding revealed failure of the resin, while failure occurred at the interface between the resin and PDMS membrane when measured after 1 week. This indicates that failure occurred within the resin until the resin was sufficiently cured and this is why the bonding strengths in air, DI water and PBS were similar; 1 week after bonding, the bonding strengths achieved by bonding in air and using BiLT were comparable.

The developed BiLT enables complete liquid filling with various membranes. Many species of liquids can be encapsulated using BiLT, unless the liquids dissolve the UV resin. This feature is crucial in manufacturing HDAM and sensors, which will be introduced in section 5. Complete filling can be achieved by direct deposition of a thin film, as introduced in section 4; however, this process can only be used to encapsulate non-volatile liquids, and the type of sealing membrane is also limited. The major drawback of this technology is that the substrate must be UV-transparent and the device should not contain UV-sensitive materials. For example, dye-sensitized photovoltaic cells, which are employed as transparent solar cells and optical sensors, require the encapsulation of electrolytes. However, BiLT cannot be used for encapsulation, because the cells have dyes that degrade after being exposed to UV.

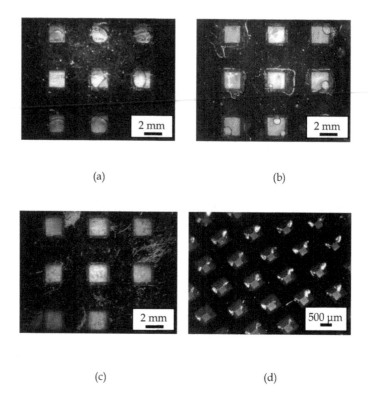

Figure 4. Bonding results for resin thicknesses of (a) 27, (b) 17 and (c) 8.1 μm. Red-dyed water was encapsulated into the cavities. The transparent parts in the cavities are excess resin. When a certain amount of resin was used, excess UV resin or air was found in the cavities, as shown in (d).

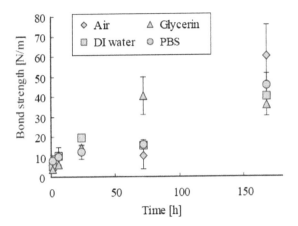

Figure 5. Bond strength after BiLT in different media as a function of time.

4. Liquid encapsulation by direct deposition of a thin film

Thin film deposition onto a solid is typically conducted under vacuum. Some liquids, such as silicone oil and ionic liquids, have extremely low vapor pressure and do not evaporate under vacuum. A thin film of metal or polymer can be directly deposited onto such low-vapor-pressure liquids. A thin silver film was deposited onto an ionic liquid to manufacture a mirror for a space telescope [33].

Parylene, or poly(para-xylylene), is widely used in fields of MEMS and microTAS due to favorable characteristics, such as transparency, mechanical strength (3.2 GPa), biocompatibility, gas sealing efficacy, and has the ability to be conformally coated using chemical vapor deposition [34-36]. Parylene has been used to form a microspring with a low spring constant [37] and as a substrate and/or a protective layer for the manufacture of microelectrodes [38]. In addition, the surfaces of PDMS microchannels have been coated with parylene to protect against protein adsorption [39].

The typical pressure for parylene deposition is several pascals. Therefore, liquids with vapor pressures less than this can remain in the liquid phase during the parylene deposition process. Binh-Khiem et al. proposed parylene on liquid deposition (POLD), where parylene is directly deposited onto a low-vapor-pressure liquid, such as silicone oil [34-36]. A feature of liquids is that sufficiently small droplets can have perfectly spherical shape due to surface tension; however, when the droplets are not small, the shape is deformed by gravity. Spherical droplets can be used as lenses. The focal length of a lens is in the same order as the lens diameter; therefore, small lenses enable compact optical systems. POLD can be used to form spherical microlens arrays; transparent liquid can be directly deposited onto silicone oil droplets. A film is formed at the droplet surface, so that no air is included in the lens, i.e., a filling rate of 100%

is achieved. Parylene is more flexible than metals; therefore, by integrating electrodes beneath the liquid lens coated with parylene, the lens can be deformed to vary the focal length according to the voltage applied to the electrodes [34].

The direct deposition on liquid approach enables perfect liquid encapsulation with good reproducibility. The useful characteristics of parylene or specific metals can be exploited; however, the disadvantage is that the liquids to be encapsulated must be non-volatile and the type of sealing material is limited. For example, this approach cannot be applied to HDAM, because it is preferable to seal the liquid with flexible membranes.

5. Applications of liquid encapsulation technology

5.1. Drug delivery

In DDS, the release of medicine is designed so that the drug efficacy is high. MEMS-based DDS are expected to convey medicine to the vicinity of diseased sites by exploiting the small MEMS size and drug release occurs when appropriate. Santini et al. proposed the controlled release of drugs from cavities that were sealed with a thin gold layer [2]. The thin gold layer is used as an anode and an electrochemical reaction occurs, with subsequent dissolution of the gold film and release of the encapsulated medicine. Drug delivery is thus controlled by electrical control of the reaction.

In this application, the fill and seal approach is employed to maintain the quality of the medicine as a priority. The package chambers are prepared with a thin gold layer as the bottom surface. Drugs are dispensed into the chambers and then sealed with a water-proof membrane using epoxy adhesive. Details of this process can be found in the literature [2].

5.2. Hydraulic amplification

Some MEMS applications require large displacement of several tens of micrometers. High-flow-rate microvalves are one such application. To satisfy the requirements, hydraulic amplification has been reported that utilizes incompressible fluid in a microchamber with an input surface that is larger than the output surface [17-20]. Another example is a tactile display [21,22]; Figure 6 shows an illustration of an array of MEMS actuators that mechanically deform the fingertip and stimulate tactile receptors. These receptors typically require skin deformation of several tens of micrometers, which could be facilitated by hydraulic amplification. Tactile displays are developed to offer a new approach to human-machine interface for virtual reality applications and interactive devices such as pointing devices or game controllers, and also to support visually impaired persons. In the microvalve applications, a working fluid can be used as an incompressible fluid for hydraulic amplification. However, tactile displays cannot afford to have an external fluidic system to drive a working fluid; therefore, complete encapsulation of the liquid is necessary. In addition, the sealing membranes must be flexible. HDAMs have been successfully developed by encapsulating incompressible and non-volatile glycerin in microchambers with flexible and largely deformable PDMS membranes (mixture of DC 3145

CLEAR and RTV thinner, Dow Corning Toray Inc.) via UV curable resin using BiLT. Figure 2 shows a schematic of a HDAM [21,22].

Figure 6. Tactile display that employs an array of MEMS actuators.

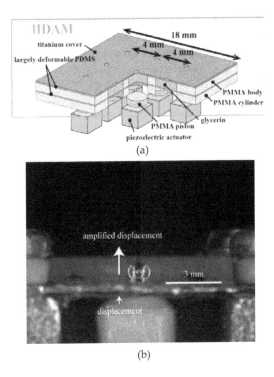

Figure 7. Large displacement MEMS actuator array that consists of HDAM with glycerin-filled cavities prepared using BiLT, and piezoelectric actuators. (a) Schematic image of HDAM, and (b) micrograph showing the large deformation of a PDMS membrane.

The HDAM shown in Figure 7(b) was combined with piezoelectric actuators and applied to develop large-displacement MEMS-actuators, with a particular aim to application in MEMS-based tactile displays [21,22,40]. When applied to a vibrational Braille code display, it was experimentally verified that the large-displacement HDAM could display Braille codes more efficiently than a static display. This is because both fast and slow adaptive tactile receptors could be used to detect the displayed patterns when individual cells were vibrated at several tens of hertz [40]. When the actuation of the large displacement MEMS actuators was controlled both spatially and temporary, different surface textures, such as rough and smooth, could be displayed.

5.3. Flexible capacitive sensor

Highly sensitive pressure sensors are expected to be applied in humanoid robots and medical instruments to detect tactile sensation, which would enable safe physical interaction with the environment, including human contact. MEMS-based capacitive sensors that have simple structures composed of electrodes and a dielectric component have been widely studied, due to good compatibility with MEMS fabrication technologies. Capacitive sensors require not only high sensitivity, but also flexibility to detect the pressure applied to curved surfaces. Silicon-based MEMS capacitive sensors have been developed; however, silicon is brittle, which makes it difficult for the sensors to conform to a curved surface. Therefore, polymer-based flexible sensors have been proposed and demonstrated. Polymer-based flexible sensors are typically used to maintain flexibility with air as the dielectric; however, air has a relatively low dielectric constant. A solid dielectric may enhance the sensitivity, but impairs the flexibility of the sensor. Therefore, a polymer-based capacitive sensor that uses a dielectric liquid has been proposed, as depicted in Figure 8 [23]. DI water and glycerin have high relative dielectric constants of approximately 80.4 and 47; therefore, the proposed sensor with such liquids can have high sensitivity while maintaining flexibility. The capacitance of the electrodes increases when pressure is applied to the device. PDMS is used as a structural material in this device. An escape reservoir is designed to allow an incompressible liquid, such as DI water and glycerin, to move from the cavity between the electrodes when pressure is applied to the sensor, which allows the flexible sensors to deform and vary the capacitance. The proposed microsensor has been fabricated, and both high sensitivity and flexibility have been experimentally demonstrated.

5.4. Dye-sensitized photovoltaic cell

Dye-sensitized photovoltaic cells are currently attracting widespread scientific and techno-logical interest as a high efficiency, low-cost, and transparent alternative to inorganic solar cells. Figure 9 shows a schematic illustration of the structure and operation principle of the dye-sensitized photovoltaic device. The cell consists of two electrodes and an encapsulated liquid electrolyte that contains iodide and triiodide ions. The cathode is a highly porous nanocrystalline semi-conductive titanium dioxide (TiO_2) layer, in many cases consisting of TiO_2 nanoparticles, deposited on a transparent electrically conductive glass. TiO_2 absorbs only UV light; therefore, dye is adsorbed onto the TiO_2 layer to utilize the light with a wider range of wavelength. The counter electrode (anode) is a transparent electrically conductive glass with a platinum catalyst. The device is transparent and is colored according to the dye employed.

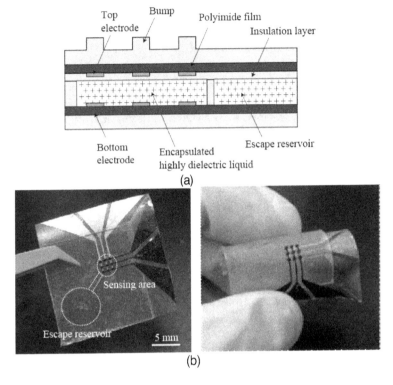

Figure 8. a) Cross-sectional view of a capacitive sensor with encapsulated liquid dielectric to enhance the sensitivity while maintaining flexibility. (b) Fabricated capacitive sensor.

When light passes through the electrically conductive transparent glass electrode, the dye molecules are excited and transfer an electron to the semiconducting TiO_2 layer via electron injection. The electron is then transported through the TiO_2 layer and collected by the conductive layer on the glass. The mediator (I-/I3-) undergoes oxidation and regeneration in the electrolyte. Electrons lost by the dye molecules to the TiO_2 layer are replaced by electrons from the iodide and triiodide ions in the electrolyte, thereby generating iodine or triiodide, which in turn obtains electrons at the counter electrode, culminating in a current flow through the external electrical load. This is the mechanism for the conversion of light energy received by the device to electricity [41]. This device has an interesting feature in that it reacts strongly to light that enters through the TiO_2 layer.

The dye-sensitized photovoltaic cell has been conventionally studied as a solar cell, where miniaturization was not considered. However, when the cells are microfabricated and arrayed, they can be used as a transparent optical sensor. Shigeoka et al. proposed to microfabricate a transparent optical sensor on eyeglasses, which could detect the pupil position by detecting reflection from the eye, as shown in Figure 10 [42]. The light reflected from the pupil is considered to be smaller than that from the white. The sensor reacts strongly to light from the TiO_2 electrode side, i.e., when the TiO_2 layer electrode is faced towards the eyes, it detects only the light reflected from the pupil and white of the eye, without being affected by the light incident on the device from the environment.

Figure 11 shows a schematic of the processes used for fabrication of this device. The most critical part is encapsulation of the electrolyte. The conductive layer (ITO) is firstly patterned on the glass substrate using photolithography. TiO_2 nanoparticles are patterned on the cathode using a lift off process. The device is subsequently annealed in air at 450 °C for 60 min and then dipped in a ruthenium-containing dye solution for 60 min to ensure the dye is adsorbed onto the TiO_2 nanoparticles. The two glass substrates are bonded via a hot melt film and application of 600-800 kPa at 100 °C. Lastly, the liquid electrolyte is flowed from the inlet hole into a channel between the two electrodes, and then the inlet and outlet holes are covered by end seals. The dyes used are UV sensitive; therefore, BiLT was not applicable to this liquid encapsulation process, and the fill and seal approach was used instead. However, the filling rate of the electrolyte was quite high and no interfusion of air between the electrodes was observed. An array of the dye-sensitized photovoltaic devices successfully detected the pupil position. The line-of-sight (LOS) was successfully deduced [42, 43] from the obtained pupil position and the front image of the subject, acquired using a CCD camera on the eyeglasses.

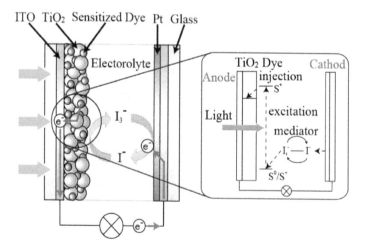

Figure 9. Structure and operation principle of the dye-sensitized photovoltaic device.

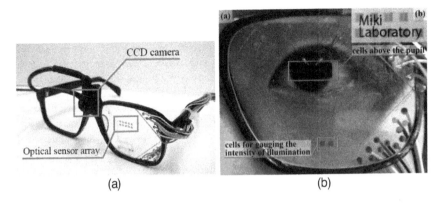

(a)   (b)

Figure 10. a) Array of dye sensitized photovoltaic cells patterned onto eyeglasses to detect the pupil position. The electrolyte was encapsulated between the electrodes. (b) Photograph of the sensor when worn by a subject.

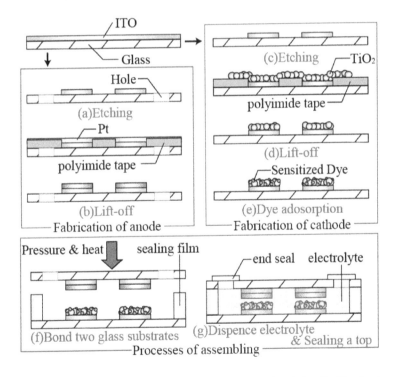

Figure 11. Fabrication process to produce an array of dye-sensitized photovoltaic devices. The fill and seal approach was employed to encapsulate the electrolyte (g).

6. Conclusion

This chapter has reviewed liquid encapsulation technologies and their applications to the manufacture of innovative MEMS devices that exploit the useful characteristics of liquid; liquid is deformable and liquid droplets form a perfectly spherical shape by surface tension. Other liquids have high relative dielectric constants. Liquids can be used as drugs for DDS and fuels for power MEMS. Appropriate liquid encapsulation technologies must be selected according to the liquid to be encapsulated. The fill and seal approach, bonding-in-liquid technique, and direct deposition of a thin film were discussed in this chapter, all of which have both advantages and disadvantages.

The use of liquid in MEMS packaging is quite a new technology. The author is convinced that more and more liquid encapsulation technologies will be developed and contribute to the further development of innovative liquid-encapsulating MEMS devices.

Acknowledgements

This work was supported by a Grant-in-Aid for Young Scientists (B) (21760202) from the Ministry of Education, Culture, Sports, Science and Technology (MEXT) of Japan, the Strategic Information and Communications R&D Promotion Programme (SCOPE) (092103005) of the Japan Ministry of Internal Affairs and Communications (MIC), and the Information Environment and Humans research area of PRESTO (Precursory Research for Embryonic Science and Technology) from the Japan Science and Technology Agency (JST).

Author details

Norihisa Miki

Address all correspondence to: miki@mech.keio.ac.jp

Department of Mechanical Engineering, Keio University, Hiyoshi, Kohoku-ku, Yokohama, Kanagawa, Japan

References

[1] Senturia, S. D. Microsystem Design. Kluwer Academic Publishers; (2001).

[2] Santini, J. T, Cima, M. J, & Langer, R. A controlled-release microchip. Nature (1999). , 397, 335-338.

[3] Santini, J. T, Richards, A. C, Scheidt, R, Cima, M. J, & Langer, R. Microchips as con-
     trolled drug-delivery devices. Angewandte Chemie International Edition (2000)., 39,
     2396-2407.

[4] Li, Y, Shawgo, R. S, Tyler, B, Henderson, P. T, Vogel, J. S, Roesnberg, A, Storm, P. B,
     Langer, R, Brem, H, & Cima, M. J. In vivo release from a drug delivery MEMS de-
     vice. Journal of Controlled Release (2004)., 100, 211-219.

[5] Richards Grayson ACShawgo RS, Li Y, Cima MJ. Electronic MEMS for triggered de-
     livery. Advanced Drug Delivery Reviews (2004)., 56, 173-184.

[6] Epstein, A. H, & Senturia, S. D. Macro power from micro machinery. Science (1997).

[7] Miki, N, Teo, C. J, Ho, L. C, & Zhang, X. Enhancement of rotordynamic performance
     of high-speed micro-rotors for power MEMS applications by precision deep reactive
     ion etching. Sensors and Actuators A: Physical (2003)., 104, 263-267.

[8] Holladay, J. D, Jones, E. O, Phelps, M, & Hu, J. L. Microfuel processor for use in a
     miniature power supply. Journal of Power Sources (2002).

[9] Arana, L. R, Schaevitz, S. B, Franz, A. J, Schmidt, M. A, & Jensen, K. F. Journal of Mi-
     croelectromechanical Systems (2003)., 12(5), 600-612.

[10] Manz, A, Graver, N, & Widmer, H. M. Miniaturized total chemical-analysis systems-
     a novel concept for chemical sensing. Sensors and Actuators B: Chemical;, 1, 244-248.

[11] Shoji, S. Fluids for sensor systems. Topics in Current Chemistry (1998)., 194, 163-168.

[12] Huikko, K, Kostiainen, R, & Kotiaho, T. Introduction to micro-analytical systems: bi-
     oanalytical and pharmaceutical applications. European journal of pharmaceutical sci-
     ences (2003)., 20, 149-171.

[13] Ota, H, Kodama, T, & Miki, N. Rapid formation of size-controlled three-dimensional
     hetero-cell aggregates using micro-rotation flow for spheroid study. Biomicrofluidics
     (2011).

[14] Gu, Y, & Miki, N. Multilayered microfilter using a nanoporous PES membrane and
     applicable as the dialyzer of a wearable artificial kidney. Journal of Micromechanics
     and Microengineering (2009).

[15] Yokota, S, Kawamura, K, Takemura, K, & Edamura, K. High-integration micromotor
     using electro-conjugate fluid (ECF). Journal of Robotics and Mechatronics (2005).,
     17(2), 142-148.

[16] Yamaguchi, A, Takemura, K, Yokota, S, & Edamura, K. A robot hand using electro-
     conjugate fluid. Sensors and Actuators A: Physical (2011)., 170, 139-146.

[17] Roberts, D. C, Hanqing, L, Steyn, J. L, Yaglioglu, O, Spearing, S. M, Schmidt, M. A, &
     Hagood, N. W. A piezoelectric microvalve for compact high-frequency, high-differ-

ential pressure hydraulic micropumping systems. Journal of Microelectromechanical Systems (2003). , 12(1), 81-92.

[18] Kim, H, & Najafi, K. Electrostatic hydraulic three-way gas microvalve for high-pressure applications. Proceedings of the 12th International Conference on Miniaturized Systems for Chemistry and Life Sciences, MicroTAS October (2008). San Diego, USA., 2008, 12-16.

[19] Kim, H, & Najafi, K. An electrically-driven large-deflection high-force, micro piston hydraulic actuator array for large-scale microfluidic systems. Proceedings of 22th IEEE International Conference on Micro Electro Mechanical Systems, MEMS January (2009). Sorrento, Italy., 2009, 25-29.

[20] Wu, X, Kim, S. H, Ji, C. H, & Allen, M. G. A piezoelectrically driven high flow rate axial polymer microvalve with solid hydraulic amplification. Proceedings of 21st IEEE International Conference on Micro Electro Mechanical Systems, MEMS January (2008). Tuscon, USA., 2008, 13-17.

[21] Arouette, X, Matsumoto, Y, Ninomiya, T, Okayama, Y, & Miki, N. Dynamic characteristics of a hydraulic amplification mechanism for large displacement actuators systems. Sensors (2010). , 10, 2946-2956.

[22] Ninomiya, T, Okayama, Y, Matsumoto, Y, Arouette, X, Osawa, K, & Miki, N. MEMS-based hydraulic displacement amplification mechanism with completely encapsulated liquid. Sensors and Actuators A: Physical; , 166, 277-282.

[23] Hotta, Y, Zhang, Y, & Miki, N. A flexible capacitive sensor with encapsulated liquids as dielectric. Micromachines (2012). , 3, 137-149.

[24] Tong, Q. Y, & Gosele, U. Semiconductor wafer bonding science and technology: John Wiley & Sons, Inc.; (1999).

[25] Miki, N. Wafer bonding techniques for MEMS. Sensor Letters (2005). , 3(4), 263-273.

[26] Miki, N, & Spearing, S. M. Effect of nanoscale surface roughness on the bonding energy of direct-bonded silicon wafers. Journal of Applied Physics (2003). , 94, 6800-6806.

[27] Niklaus, F, Stemme, G, Lu, J. Q, & Gutmann, R. J. Adhesive wafer bonding. Journal of Applied Physics (2006).

[28] Dlutowski, J, Biver, C. J, Wang, W, Knighton, S, Bumgarner, J, Langebrake, L, Moreno, W, & Cardenas-valencia, A. M. The development of BCB-sealed galvanic cells. Case study: aluminum-platinum cells activated with sodium hypochlorite electrolyte solution. Journal of Micromechanics and Microengineering (2007). , 17, 1737-1745.

[29] Okayama, Y, Nakahara, K, Xavier, A, Ninomiyia, T, Matsumoto, Y, Hotta, A, Omiya, M, & Miki, N. Characterization of a bonding-in-liquid technique for liquid encapsulation into MEMS devices. Journal of Micromechanics and Microengineering (2010).

[30] Zhang, Y, Ishida, M, Kazoe, Y, Sato, Y, & Miki, N. Water vapour permeability control of PDMS by the dispersion of collagen poweder. TEEE: Transactions on Electrical and Electronic Engineering (2009). , 4(3), 442-449.

[31] Sawano, S, Naka, K, Werber, A, Zappe, H, & Konishi, S. Sealing method of PDMS as elastic material for MEMS. Proceedings of 21st IEEE International Conference on Micro Electro Mechanical Systems, MEMS January (2008). Tuscon, USA., 2008, 13-17.

[32] Antelius, M, Fischer, A. C, Niklaus, F, Stemme, G, & Roxhed, N. Hermetic integration of liquids using high-speed stud bump bonding for cavity sealing at the wafer level. Journal of Micromechanics and Microengineering (2012).

[33] Borra, E. F, Seddiki, O, Angel, R, Eisenstein, D, Hickson, P, Seddon, K. R, & Worden, S. P. Deposition of metal films on an ionic liquid as a basis for a lunar telescope. Nature (2007). , 447(7147), 979-981.

[34] Binh-khiem, N, Matsumoto, K, & Shimoyama, I. Polymer thin film deposited on liquid for varifocal encapsulated liquid lenses. Applied Physics Letters (2008).

[35] Binh-khiem, N, Matsumoto, K, & Shimoyama, I. Tensile film stress of parylene deposited on liquid. Langmuir (2010). , 26(24), 18771-18775.

[36] Takamatsu, S, Takano, H, Binh-khiem, N, Takahata, T, Iwase, E, Matsumoto, K, & Shimoayma, I. Liquid-phase packaging of a glucose oxidase solution with parylene direct encapsulation and an ultraviolet curing adhesive cover for glucose sensors. Sensors (2010). , 10(6), 5888-5898.

[37] Suzuki, Y, & Tai, Y. C. Micromachines high-aspect-ratio paryelene spring and its application to low-frequency accelerometers. Journal of Microelectromechanical Systems (2006). , 15(5), 1364-1370.

[38] Takeuchi, S, Ziegler, D, Yoshida, Y, Mabuchi, K, & Suzuki, T. A parylene flexible neural probe integrated with micro fluidic channels. Lab on a Chip (2005). , 5, 519-523.

[39] Sasaki, H, Onoe, H, Osaki, H, Kawano, R, & Takeuchi, S. Parylene-coating in PDMS microfluidic channels prevents the absorption of fluorescent dyes. Sensors and Actuators B: Chemical (2010). , 150(1), 478-482.

[40] Watanabe, J, Ishikawa, H, Arouette, X, Matsumoto, Y, & Miki, N. Demonstration of vibrational Braille code display using large displacement micro-electro-mechanical system actuators. Japanese Journal of Applied Physics (2012). FL11.

[41] Regan, O, & Gratzel, B. M. A low cost, high-efficieny solar cell based on dye-sensitizd colloidal TiO_2 films. Nature (1991).

[42] Shigeoka, T, Muro, T, Ninomiya, T, & Miki, N. Wearable pupil position detection system utilizing dye-sensitized photovoltaic devise. Sensors and Actuators A: Physical (2008).

[43] Oikawa, A, & Miki, N. MEMS-based eyeglass type wearable line-of-sight detection system. Proceedings of 2011 IEEE International Conference on Robotics and Automation, ICRA May (2011). Shanghai, China., 2011, 9-13.

Chapter 2

Electroporation Based Drug Delivery and Its Applications

Tuhin Subhra Santra, Pen-Cheng Wang and
Fang Gang Tseng

Additional information is available at the end of the chapter

1. Introduction

When a certain strong electrical pulse applied across a cell or tissue, the structures of the cell or tissue would be rearranged to cause the permeabilization of the cell membrane, named in early 1980's "electroporation"[1]. The theoretical and experimental studies of electric field effects on living cells with their bilayer lipid membrane has been studies in 1960's to 1970's century [1-6]. During these years, the researches were primarily dealt with reversible and irreversible membrane breakdown in vitro. Based on these research, the first gene transfer by custom-built electroporation chamber on murine cells was performed by Neumann et al. in 1982 [7]. When electric field (E≈0.2V, Usually 0.5-1V) applied across the cell membrane, a significant amount of electrical conductivity can increase on the cell plasma membrane. As a result, this electric field can create primary membrane "nanopores" with minimum 1 nm radius, which can transport small amount of ions such as Na^+ and Cl^- through this membrane "nanopores". The essential features of electroporation included (a) short electric pulse application (b) lipid bilayer charging (c) structural rearrangements within the cell membrane (d) water-filled membrane structures, which can perforate the membrane ("aqueous pathways" or pores) and (e) increment of molecular and ionic transportation [8]. In conventional electroporation (Bulk electroporation) technique, an external high electric field pulses were applied to millions of cells in suspension together in-between two large electrodes. When this electric field was above the critical breakdown potential of the cell, a strong polarization of the cell membrane occur due to the high external electric field. Applying a very high electric field could be resulted in the formation of millions of pores into the cell membrane simultaneously without reversibility [9]. Several methods other than electroporation can be used for gene transfer like microprecipitates, microinjection, sonoporation, endocytosis, liposomes, and biological vectors [10-16]. But electroporation have some advantages when compared to other gene transfer methods such as, (a) easy and rapid operation with high reproducibility due to control of electrical parameters (b) higher transformation efficiency when compared to $CaCl_2$ and PEG mediated chemical transformation (c) controllable pore size with variation of electrical pulse and minimizing effect of cytosolic components, and (d) easy to uptake DNA into cells with smaller amount, when compared to other techniques [17-19]. For bulk electroporation, drug delivery can be performed in homogeneous electric field, whereas as single cell electroporation (SCEP), can introduce an inhomogeneous electric field focused on targeted single adherent or suspended cell without affecting other neighboring cells. Both techniques can deliver molecules such as DNA, RNA, anticancer drugs into cells in–vitro and in-vivo. However SCEP is more advanced technique compared to the bulk electroporation technique. Recently researchers are concentrating on more advanced research area, such as localized single cell membrane electroporation (LSCMEP), which is an efficient and fast method to deliver drugs into single cell by selective and localized way from millions of cells. This LSCMEP can judge cell to cell variation precisely with their organelles and intracellular biochemical effect. This process can deliver more controllable drug delivery inside the single cell with application of different pulse duration. Both single cell electroporation (SCEP) and localized single cell membrane electroporation (LSCMEP) can provide high cell viability rate, high transfection efficiency, lower sample contamination, and smaller Joule heating effect in comparison with bulk electroporation (BEP) process.

2. Electroporation conditions

To achieve excellent gene delivery into the cells, several electroporation conditions will be accomplished during experiment. Also these electroporation conditions depend upon cell to cell variation. Generally these conditions can be divided into three categories (a) cellular factors (b) physiochemical factors and (c) electrical parameters.

2.1. Cellular factors

The gene delivery by electroporation into living cells should take place with high transfection efficiency and high cells viability in a physiological unperturbed state, so that, the effect of gene on a specific cellular function can be measured. The transformation efficiency can be influenced by growth phase of the cells, cell density, cell diameter, cell rigidity etc. The growth period of the cells in higher transformation success can be achieved from early to middle phase [20]. For electroporation, two main parameters needed to be optimized, one is electric field strength and the other is the pulse duration of electric field. When we apply voltage between two electrodes (this two electrodes maintain some distance), the pulse is generally an exponentially decayed signal with a time constant given by the product of the capacitance and resistance of the buffer solution. For any kind of field strength and pulse duration, the extent of macromolecular entry and degree of mortality will vary among different cell lines [21]. If transmembrane potential (TMP) difference is proportional to the cell size, the electric field strength will be more sensitive for larger cells compared to smaller cells [22]. Also it has been reported that, transmembrane potential difference is related to cell angles and the directions of applied electric fields, where TMP values linearly proportional to the external applied electric field and cell diameter [23]. For the detection of specific effect of electroporated antibody, cellular function can depends on many variables, such as (a) concentration and affinity of introduced antibody into the target cell (b) restriction of antibodies to bind by target molecules (c) antibody can effect by intracellular concentration of target molecules (d) target molecules cellular factor such as epitopes(s) which can recognize the antibodies are unable to bind with target molecules (e) the cellular distribution of target molecules is accessible or not for antibody [21]. The cell viability during electroporation is also an important factor. Several literatures reported that nucleic acid molecules can be delivered in a highly efficient manner by optimizing the electroporation parameters, and the optimized electroporation conditions using a fluorescently labeled transfection control siRNA resulted in 75% transfection efficiency for Neuro-2A, 93% for human primary fibroblasts, and 94% for HUVEC cells, as analyzed by flow cytometry [24]. Saunders et al. have shown the successful uptake of trypan blue and FDA in cells, protoplasts and pollen from different plants using variety of pulse generator for optimizing the electroporation conditions [25].

2.2. Physiochemical factors

Physiochemical factors are more important for electroporation. This phenomena can occur during tissue development which contain the transportation, consumption of nutrients and oxygen, waste generated by cells, mechanical loading of tissue or cells, electromechanical phenomena (piezoelectricity), chemomachanical phenomena (swelling), electrochemical phenomena (Debye length) or osmotic phenomena (transport through the cell membrane). During cell culture stage, cells have to proliferate, colonize homogeneously in porous scaffolds and synthesized extracellular matrixes [26]. Different type of molecules or elements can interact with cells during cell culture [27]. Among all of the soluble elements, oxygen molecules possess the major importance for tissue growth particularly for osteoarticular system [28-29]. The magnitude of cell local oxygen consumption could be affected by cell concentration and temperature. The oxygen molecules passes through the cell membrane subject to enzymatic chemical, which is familiar as fundamental in enzymology. The oxygen consumption (R_s) per unit area of cell layer with surface density σ_{cell} can be described as the following expression

$$R_S = \sigma_{cel} \times V_{max} \times \frac{C}{C + K_M} = -R_{max} \times \frac{C}{C + K_M} \tag{1}$$

Where KM is the Michaelis constant, C is the nutrient molecular concentration and the negative sign indicated that all cell layers have a sink effect. The maximum oxygen consumption rates V_{max} depends upon cell types and it can vary several order of magnitude. The oxygen consumption in unit volume such as porous substrate as written as

$$R_V = S_V \times R_S \tag{2}$$

This law also can be utilized for other biological phenomena such as cell population growth, drug uptake by tumor cells or absorption of biochemical molecules within kidney [26]. The electroporation efficiency can be affected by ionic composition of buffer solution. The resistivity and RC time constant of the electric pulse can be determined by ionic concentration of the buffer as written as [20]

$$V = V_0 \exp(\frac{-t}{RC}) \tag{3}$$

$$\tau = RC \tag{4}$$

where, V is the voltage across the pulsing chamber, V_0 is initial voltage, t is the time after starting of the pulse, R is the resistance of suspension, C is the capacitor of the capacitance, and τ is the time constant. The salt concentration of the electroporation buffer as well as pH of the buffer solution can affect the electroporation efficiency [30-31]. Generally the pH value 7.2 can be considered as an appropriate value for electroporation condition. The permeability of the cell membrane depends upon the solubility properties (such as salt composition, pH), charges or chemistry and solute size. The water molecule can transport inside and outside by osmotic balance. Osmosis can maintain the turgor pressure of the cells, across the cell membrane between the cell interior and relatively hypotonic environment [32]. The swelling properties of biological tissues can be explained by osmotic disjoining pressure [33]. Also the electroporation efficiency is much better, by introducing gene into cells at (0-4 °C) compared to elevated temperature during electroporation experiment [34-35]. This low temperature helps to protect the rapid resealing of the pores and enhance the uptake efficiency of gene inside the cell [17]. It has been reported that high transformation efficiency can be achieved by cell suspension of slow growing mycobacteria at elevated temperature [36]. Regarding the transfer of DNA into cells , it has been shown that cooling at the time of permeabilization and subsequent incubation (37 °C), can enhance the transformation efficiencies and cell viability [37]. Some of the authors has reported that, the use of low conductivity medium for DNA transfer, can increase the cell viability and transformation efficiency [37]. Increasing the amount of DNA into the pulse chamber can increase linearly transfection level [38-39]. However the toxic effect can be observed for high DNA concentration [39-40]. It is generally considered the use of calcium in the medium during electroporation for not causing high intracellular level of electrolyte. However some researchers use calcium and magnesium into the buffer solution for performing DNA transformation into the cell. In such a condition, DNA with calcium ions can act as positively charged 'glue' and attracted by the negatively charged ions on the exterior cell membrane, as a result, DNA molecules are approximating to the membrane before the electroporation process [41-42].

2.3. Electrical parameters

Electrical parameters are the most important factors to achieve high transformation efficiency and high cell viability during successful gene transfer into living cells. The electrical parameters mainly depend upon electric field strength, pulse length, number of pulses, time between two pulses and etc. Cell plasma membrane always have a tendency to protect the cytoplasmic volume from outside of any exogenous molecules. Cell membrane also continuously prevent cell to cell fusion. However, if we apply external electric field pulses and if this electric field just surpasses the capacitance of the cell membrane, then transient electropermeabilized state can occur, which allow the delivery of various extracellular molecules, such as drugs, antibodies, DNA, RNA, dyes, tracers and oligonucleotides from outside of the cell to inside of the cell. If the molecular size is small, it can enter inside the cell membrane by diffusion after electropermeabilization. However if the size is large, the molecules can enter into the cell through electrophoretically driven process as like DNA transferring into the cell membrane. Previously it was reported that, short and strong electric field pulses can make the membrane permeable in a spontaneously reversible way [43]. Also, it was reported, an extremely short pulse in nanosecond range with very high voltages, cellular organelles can be electroporated without cell membrane permeabilization [44]. The cell membrane permeabilization area can be controlled by pulse amplitude. By this permeabilization area, diffusion can take place into the cell membrane [45]. The degree of permeabilization can be controlled by the pulse duration and pulse number, where the longer the pulse, the greater the perturbation of the membrane in a given area [46]. Also it has been reported that area of the membrane being permeabilized is larger on the pole facing positive electrode, but degree of permeabilization is greater on the cell, where pole facing negative electrode [47]. However high transformation efficiency can be obtained, when three successive pulses with two intermittent cooling steps of one minute in each or single pulse without cooling for transformation of *Enterococcus faealits*, *E. coli* and *Pseudomonas putida* [38]. Kinetic study of electropermeabilization leads to 5 steps.

Time dependent electropermeabilization		
Trigger	The external field induces an increase in the transmembrane potential up to the critical permeabilizing threshold	μm
Expansion	A Time dependent membrane transition occurs as long as the field is maintained at a overcritical value	ms
Stabilisation	A dramatic recovery of the membrane organization take place as soon as the field is subcritical	ms
Resealing	The annihilation of leaks is slow	s
Memory	Cell viability is preserved but membrane structural (flip flop) and physiological properties (macropinocytosis) recover on a much longer time	h

Table 1. Time dependence of electropermeabilization. Permission to reprint obtained from Elsevier [50].

Table-1, illustrates the five steps where "Induction step" describes the field induced membrane potential increase which provides local defects, when it reached to a certain critical value (above 200mV). Here mechanical strength of the cell membrane depends upon buffer composition. The "Expansion step" comes when field presents with a strength larger than a critical value. In this case electromechanical stress present. "Stabilization step" indicates, field intensity is lower than threshold value, a stabilization process will take place in a few milliseconds. As a result membrane will be permeabilized for small molecules. "Resealing step" demonstrates a slow resealing on a scale of seconds and minutes. The "Memory effect" comes due to some changes of the membrane properties for longer time, such as an hours, but cell behavior is still normal [48-50]. Table-2 demonstrate electroporation conditions of various cell types [51], where electric field strength, pulse length, no of pulses, time between two pulses vary in each different type of cells.

Cell type	Voltage (Volt)	Pulse length (µS)	Number of pulses (sec)	Time between pulses (second)	Number of cells
HMSC	700V	90	5sec	0.1	75,000
HUVEC	250V	150	-	-	75000
RPTEC	300V	300	-	0.1	75000
Human T-Cells	300V	400	-	0.1	200000
NHDF-neo	900V	70	5sec	5	75000
PC-12	450V	200	-	-	75000
Rat astrocytes	300V	90	0.1sec	0.1	75000
NHA	450V	120	0.1sec	0.1	75000
K562	350V	130	0.1sec	0.1	150000

Table 2. Electroporation conditions for various cell types. Permission to reprint obtained from RNA society [51].

3. Single cell electroporation

3.1. Prospect of SCEP over Bulk Electroporation (BEP)

For single cell electroporation (SCEP), the electric field parameters can be controlled to avoid cell death. In SCEP, where an inhomogeneous electric field is applied locally surrounding the single cell adhesion or suspension, whereas in bulk electroporation (BEP), a homogeneous electric field is applied to suspension of millions of cells together. Fig.1. shows two types of conventional bulk electroporation(BE) chamber, to apply electric field with suspension of millions of cells together for vitro experiment. Both figures has shown the cross sectional view with two metal electrodes.

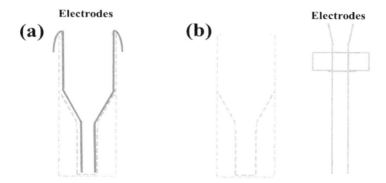

Figure 1. Bulk electroporation apparatus for vitro experiment. Two types of electroporation chamber, to apply an external electric field into the suspension of millions of cells together. Each chamber (a.b) consists cross sectional view of cuvette with two metal electrodes. Figure has redrawn with reprint permission [8].

Fig.2. demonstrates the single cell electroporation technique, where an external electric field is applied across the single cell membrane surface.

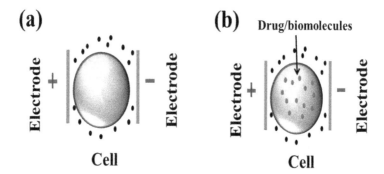

Figure 2. Single cell electroporation (SCEP) (a) Electric field was applied outside of the cell membrane (b) When external electric field reaches to a certain threshold value of the cell membrane, then cell membrane can permeabilized to deliver drug/biomolecules inside the single cell. After electroporation cell membrane reseal again.

When an external electric field beyond the certain threshold value of the cell membrane, then cell membrane can permeabilized to deliver biomolecules inside the single cell. The success rate like surviving cell for single cell electroporation is far better compared with bulk electroporation (BEP). This technique is faster and easy to perform with less toxicity and technical difficulty for application of wider tissues and cells. By this electroporation technique, the specific cell membrane region with small volume can be targeted to deliver the drugs, which can help to preserve expansive gene or molecules. Due to small volume of electroporation, different gene can be transferred in different electroporated time without cell damage. SCEP technique can provide precise temporal and spatial gene or dye delivery inside the cell. These processes are affordable methods for fluorescently labeled and genetically manipulated individual cells [52]. This level of electroporation study is more convenient to understand molecular and genetic mechanisms with their biological functions and SCEP has ability to control temporally molecular biology of the cell, which was challenging task for transgenic model systems [52]. For bulk electroporation the required voltages are very high (10^3V) and this technique has little control of individual cell resulting in suboptimal parameters [53], as a result it is difficult to achieve reversible electroporation of all cells [54]. Moreover in single cell electroporation, there is good opportunity to observe the single cell response with specific cell size, shape, status and orientation of the electric field. SCE is useful for primary culture and heterogeneous culture such as brain tissue culture [55].

The first single cell electroporation has been demonstrated by using two carbon fiber microelectrodes [46], where the electrodes (2 µm to 5 µm) was positioned from the boundary of the cell surface at an 0-20° angle and 160-180° angle with respect to the objective plane. The patch clamp technique demonstrated the single cell electroporation (SCE), where patch-clamp pipette was sealed on the cell at a 90^0 angle with respect to the microelectrodes [56]. Using this technique, from transmembrane current response, it was possible to determined electric field strength for ion permeable pore formation and kinetics of pore opening, closing as well as pore opening times [56]. The electrolyte-filled capillary (EFC) coupled with a high-voltage power supply has been used for single cell experiment [57]. For application of a large voltages across EFC, It causes the formation of pores in the cell membranes which induces an electroosmotic flow of electrolyte. Micropipettes filled with DNA or other molecules stimulated by electric field have been electroporate the single cell at the tip of the pipette and successfully deliver the molecules inside single cell [58]. Microfabricated chip was used to incorporated the biomolecules into live biological cells for single cell experiment [59]. To achieve successful single cell electroporation, cell must be isolated from its population or inhomogeneous electric field must be focused on a particular cell, leaving neighboring cells unaffected [60]. Microfabricated devices can fulfil both isolated single cell and focused the electric field on particular single cell. Also this technology can offer other functionalities into the chip. Nowadays, SCEP research is growing on rapidly for biomedical application in vivo and in vitro. However to allow selective manipulation of single organelles within a cell, the electrode size must be reduced to nanoscale level. Nanoelectrode can provide less toxicity with high cell viability during electroporation experiment. Thus the localized single cell membrane electroporation concept has come in several years [61]. Fig.3. shows the localized single cell membrane electroporation (LSCMEP) process, where electric field is applied very short region of the cell membrane.

As a result, due to permeabilization of the cell membrane, drug/biomolecules can be delivered precisely (through sub micrometer to nanometer region of the cell membrane surface) inside the single cell. By this technique selective manipulation of organelles and biochemical effects can be analyze more precisely of the individual cell and this technique have more advantage compared to SCEP. Also the cell rapture and cell death can be minimize because electric field can intense in localized region of the cell membrane compared to SCEP. But this technology is now in underdeveloped stage. Recently Boukany et al. suggested nanochannel electropo-

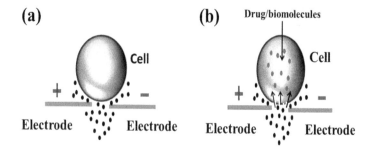

Figure 3. Localized single cell membrane electroporation (LSCMEP) technique, where drug/biomolecules can deliver precisely inside the single cell (a) Electric field was applied in a very small region of the cell membrane area (Localized way) (b) After electric field application, due to permeabilization of the cell membrane, drug/biomolecules can successfully deliver inside the single cell. Permission to reprint obtained from Springer [63].

ration with precise amount of biomolecules delivery by LSCMEP process. Where single cell has been positioned in one microchannel by optical tweezers and transfection agent was loaded to another microchannel. Two microchannel were connected by one nanochannel. Due to application of voltage between two microchannels, transfection agent was delivered through nanochannel using electrophoretically driven process and finally drugs delivered inside single cell through a very small area of the cell membrane [62]. Nawarathna et al. demonstrated localized electroporation technique using atomic force microscopy (AFM). Where modified AFM tip (0.5 μm) was used as a nanoelectrode, which was produced localized electric field into the cell membrane [61]. Fig 4.(a-h) shows the results of LSCMEP technique using AFM tip for electroporation process and Fig.4(i) demonstrated the AFM tip, which was positioned on top of the single cell for LSCMEP process.

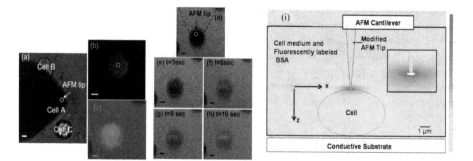

Figure 4. (a) Bright field image of AFM tip where the cell in the electroporation medium (cell A is electroporated while cell B and C are about 20 μm away from cell A). (b) Fluorescence image of rat fibroblast cell after electroporation. (c) Confocal fluorescence image of an electroporated cell. (d)-(h) Sequence of real time confocal fluorescence images of rat fibroblast cell after electroporation. (i) Calculated spatial distribution of electric field in the vicinity of the cell being electroporated. Permission to reprint obtained from American Institute of Physics (AIP) [61].

Chen et al. demonstrated localized single cell membrane electroporation (LSCMEP) by using microfluidic device. Where ITO thin film was used as microelectrode with 1 μm gap between two micro-electrodes. The ITO microelectrode with 100 nm thickness and 2 μm width intense electric field much more in between two microelectrode gap [63]. Fig.5. shows the device fabrication for localized electroporation experiment.

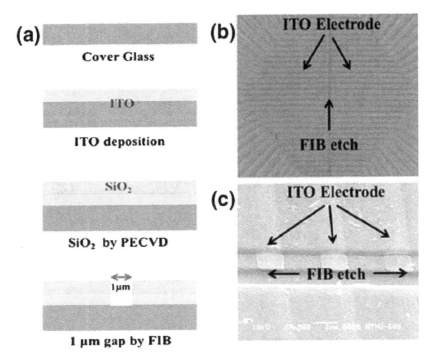

Figure 5. Fabrication process of ITO microelectrode based localized single cell electroporation chip. (a) Fabrication process step (b) Optical microscope image of patterned ITO microelectrodes. (c) SEM image of ITO microelectrodes with micro channel (FIB etch), Permission to reprint obtained from Springer [63].

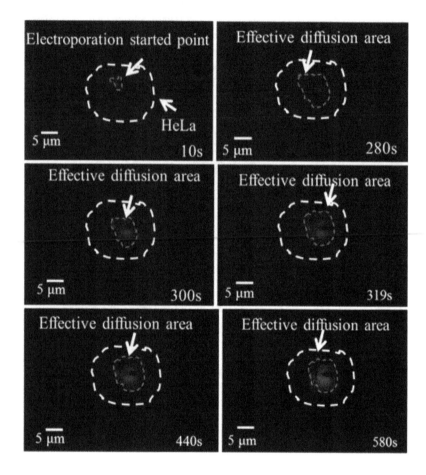

Figure 6. After application of 8Vpp 20 ms pulse, cell survival fluorescence image of HeLa cell at different time scale, Permission to reprint obtained from Springer [63].

According to the results, 0.93 μm electroporation regions were achieved successfully with 60% cell viability for 20 microsecond pulse. Fig.6. demonstrates the cell survival fluorescence image of HeLa cell at different time scale during LSCMEP process.

3.2. Pore formation on SCEP

In single cell electroporation technique, electroporation occurs in adherent cell and tissue. However single cell electroporation can be visualized for cell in suspension. In BEP, mostly the cells are in suspension as spheres, in which homogeneous electric field can be applied. But for single cell electroporation, electric field is in inhomogeneous form, which targets on a particular cell without effecting neighboring cells. Generally cell membrane described in terms of fluid mosaic membrane model [64]. Due to application of an electric electric field, the formation of pores into the cell membrane depends upon field strength with low conductance, which is approximated as electrical capacitors with infinite resistance. The pore as liquid capacitor which converts to the electrical force associated with transmembrane potential U into an expanding pressure within the aqueous pore interior [65-68]. The pore creation energy ΔE can be calculated with pressure balance by removal of planar area πr^2 and creation of a cylindrical pore edge of length $2\pi r$, can be written as

$$\Delta E = 2\Pi \gamma r - \Pi r^2 \Gamma \tag{5}$$

where surface energy approximately $\Gamma = 1 \times 10 \text{ J/m}^2$ and the edge energy approximately $\Upsilon = 1$ to 6×10^{-11} J/m [69-71]. Here Υ is constant even it is a function of r [70, 72-73]. To expand the pore radius from zero radious to r can be written as

$$\Delta E(r) = 2\Pi \gamma r - \Pi r^2 \Gamma + A / r^4 \tag{6}$$

The first term is energy related stressed pore edge with length $2\pi r$. The second term is energy to remove a circular flat lipid membrane having energy per unit area Γ and the third term is steric repulsion of the lipids with constant A. Fourth term arises when transmembrane potential V_m is nonzero, which is related to,

$$-0.5C_p V_m^2 r^2 \tag{7}$$

The transmembrane potential $\Delta\Psi_E$, in a uniform electric field E at a point M with time t can be written as

$$\Delta\psi_E(t) = \psi_{in} - \psi_{out} = -fg(\lambda)RE\cos\theta(M) \times (1 - \exp(-t / \tau)) \tag{8}$$

where f is the shape of the spheroidal cell [74] and τ is the charging time of the cell membrane, g depends upon the conductivities and R is the radius of the spherical cell. E is the field strength and $\theta(M)$ is the angle between normal to the membrane at the position M and direction of the field [55]. The exponential term can be ignored if the pulse length is longer than a few microseconds. Because induction time $\tau < 1\mu s$, the value f is generally 3/2, which is for completely insulating membrane [75].

4. Bulk electroporation

4.1. Electric field effect on cell membrane

Biological systems are mainly heterogeneous from electrical point of view [76-77]. When a high electric field pulses is applied across the cell membrane, due to rapid polarization, cell membrane can deform mechanically (e.g., suspended vesicles and cells) and is allowed to redis tribute ionic charges due to electrolyte conductivities and distributed capacitance. Initially every bilayer cell membrane structure is dielectric in nature. After application of electric field pulses, membrane conductivity can increase due to structural change of the cell membrane cause the formation of hydrophilic pores from initially formatted hydrophobic pores [78]. Generally the breakdown potential of lipid bilayer is 100-300 mv, which depends upon the lipid compositions [79]. If the pulse electric field (PEF) decreases, then breakdown voltage can increase [80-81].

To consider a cell as a sphere with a small volume of V and current is flow of charges. Both current and charges have relationship between them. If we consider the total current flow through small volume of cell V, then the current must be equal to the net flow of charges with in volume V or equal to the rate of decrease of charge with in volume or net flow of current into volume V must be accompanied by an increase of charge with in volume V. This is the principle of conservation of charge, which can be mathematically expressed as

$$I = \int_S J.ds \tag{9}$$

$$= -\frac{\partial}{\partial t}\int_V \rho dV \tag{10}$$

Using divergence theorem, $\int_S A.nds = \int_V \nabla.AdV$ then the equation can be written as

$$\int_V \nabla.JdV = -\int_V \frac{\partial \rho}{\partial t}dV \tag{11}$$

where J is the current density and ρ is the volume charge density. Now we can write equation [11] as

$$\int_V \left(\nabla.J + \frac{\partial \rho}{\partial t} \right) dV = 0 \tag{12}$$

Since equation [12] must be true irrespective of the volume, so we can write equation [12] as

$$\nabla . J + \frac{\partial \rho}{\partial t} = 0 \tag{13}$$

This is equation of continuity, which is the principle of conservation of charge where steady current involve $\frac{\partial \rho}{\partial t} = 0$ and if charges are not generated into the cell during application of electric field pulses, then $\nabla . J = 0$. Now electric field is the gradient of electric potential. So Maxwell equation becomes $\Delta^2 \psi = 0$, where Ψ denotes the electrical potential. If the conductivity of cytoplasm and external medium of the cell is higher than the cell membrane conductivity, then $\Delta \Psi$, the field induced transmembrane potential can be written as:

$$\Delta \psi = 1.5 a_{cell} E_e \cos\theta \tag{14}$$

where a_{cell} is the outer radius of the cell, E_e is the applied electric field strength and θ is angle between field line and normal to the point of interest in the membrane which can be either 0^0 or 180^0 [82-85]. Under the ideal experimental conditions like pulse width, electric field, number of pulses, removal of external electric field for resealing of the pore membrane, pulse duration and rearrangement of the membrane protein can be preserved the cell viability. If the membrane is not spherical, then equation [14] may not be right explanation. If we consider that the cell has ellipsoidal structure, then equation [14] will not be applicable. But for any practical purpose this equation can be used to evaluate the field induced transmembrane potential.

4.2. Reversible electroporation

When a strong external electric field applied across cell and tissue, then membrane conductance and permeability can increase significantly due to strong polarization of the cell membrane, as a result membrane can form nano scale defects (called nanopores). But when we switched off the external electric field, membrane can return from its conducting state to its normal state. This phenomenon is called reversible electric breakdown or reversible electroporation [86-87]. The reversible electroporation generally involves reversible electric breakdown (REB), which is generally a temporary high conducting state. This reversible electroporation influences both cell membrane as well as artificial planner bilayer lipid membrane. Reversible electroporation involve with rapid creation of many small pores, where membrane discharge occur before any critical pores can evolve from the small pores. To understand the method of electroporation of bilayer lipid membrane, it is necessary to use the method of voltage clamp [65,71,88] and charge relaxation [80,89] techniques, where for charge relaxation, kinetics of voltage decreases across the membrane after the application of short pulses (20 nsec to 10 μsec). It was also fact that originally membrane breakdown can occur before the start of membrane discharge. From the charge relaxation method, it used to show that, when membrane of oxidized cholesterol are rapidly (~ 500nsec) charged (approximately 1 V), then membrane resistance reversibly decreased by almost nine orders of magnitude [80, 89]. By this way it was first observed that reversible breakdown of planner lipid bilayer membrane and the charged could not be exceeded beyond 1.2 V, even pulse amplitude was increased further. After first electroporation, it was able to recharge again. This same phenomenon was investigated later with azolectin bilayers modified UO_2^{2+} ions and the membranes of lecithin and cholesterol in the presence of alkaloid holoturin A [90-92]. The different types of behavior of planer oxidized chelosterol membrane are shown in table III [78,80,93].

Characteristic electrical behavior	Pulse magnitude
"Reversible electrical breakdown"(REB); membrane discharge to U=0	Largest
Incomplete REB(discharge halts at U>0)	Smaller
Rupture (mechanical); slow, sigmoidal electrical discharge	Still smaller
Membrane charging without dramatic behavior on U	Smallest

Table 3. Planner bilayer membrane electroporation. Permission to reprint obtained from Elsevier [78, 80, 93].

For voltage clam method, the time resolution is 5-10 μs to monitor continuously charge at specific conductance of membrane from 10^{-8} to 10^{-1} Ω^{-1}/cm^{-2}. Thus voltage clam technique and charge relaxation technique are complement to each other [94].

4.3. Irreversible electroporation

In our earlier discussion of reversible electroporation, external electric field can permeabilize the cell membrane temporarily by which, the cell membrane can survive and the process known as "reversible electroporation" whereas, some of strong external electric field can cause t the cell membrane to permanently permeabilize (membrane becomes weak effect on conductance), by which the cell can die and the process is refer to as "irreversible electroporation". This irreversible electroporation was observed in early 1754 due to discharge of static electrical generator of the skin [95-96]. The main phenomenon of irreversible breakdown was stochastic quantities by which mean life time of membrane can abruptly decreased with increased of voltage. The pores of the bilayer membrane can be hydrophilic or hydrophobic [65]. For hydrophobic cases, the pores can be formed by hydrocarbon lipid tails. Whereas the inner surface of the pores can be covered by polar tails. The hydrophobic pores which can fill by water are energetically unfavorable [66] and thus should be short -lived. The formation of the pores during reversible electroporation can exist for longer periods of time due to hydrophilic pores. The accumulation of pores during reversible electroporation is due to membrane containing lysolecithin, which can decrease the linear tension of hydrophilic pores [97-98]. The hydrophilic pores can cause the reversible and irreversible breakdown of lipid membrane. Also every electrical field can produce the thermal effect as familiar as Joule effect is disputed, where as certain electric field is undisputed, which can provide irreversible electroporation [95]. Irreversible electroporation can affect only the membrane of living cells and spares of tissues scaffold. During irreversible electroporation, the membrane survives in two stages as (a) steady state current stage and (b) fluctuating current stage. The phenomena of irreversible electroporation can cause by charge pulse technique [80] in which membrane is charged at U=0.1 V (with pulse width 400 ns) and discharged was very slow. The large pulse of the same width, can charge the membrane towards 0.4 V, but after 300-400µs, charges can be decreased as a sigmoidal manner up to zero because of membrane rupture [78].

5. Applications

5.1. Bulk electroporation

From the last decade, the application of electroporation has been increasing rapidly. Nowadays, the electroporation technique can be applied in many way to deliver drugs, antibodies, oligonucleotides, proteins, RNA, DNA and plasmid in vivo for clinical, biotechnological and biomedical applications [42, 99-101]. Table 4 described details about gene transfer by electroporation technique with the variation of molecules/gene, targeted cells, different types of electric pulses [7, 102-117].

Ref.	Year	Authors	Recipient cells	Plasmid /gene	Pulsing CD:E_0;τ	Results
7	1982	Neumann et al.	Mouse L tk Fibroblast cells	pAGO with tk gene herpes virus(HAT select.)	CD:3×8 KV/CM;5µs 20°C, 10 min postincub., HBS (without Mg²⁺)	Sharp optimum in field strength, incubation after pulse necessary, linear plasmid better than circular, 100col./10⁶ cells/µg DNA
102	1983	Shivarova et al.	Bacillus cereus protoplasts	pUB110 from B. thuringiensis(kn resistance)	CD: 3× 14 KV/cm; 5 µsec40% PEG present, 20°C, 10 min incub.	Small objects, high electric field strength necessary, 10-fold increase in stable transformation.
103	1984	Falkner et al.	Mouse lymphoid cell lines	Plasmid with Ig κ gene	CD: 3×8 KV/cm; 5µsec 20°C, 10 min incub., DME medium + 20 mM MgCl₂ (plastic cuvette)	Two to five copies of plasmid per genome integrated in transformed clones.
104	1984	Potter et al.	Mouse B and T lymphocytes and fibroblasts	Mouse and human Ig κ gene	Pulse: ISCO 494 power supply directly discharged through cuvette, no definite pulse parameters given, estimated: 320 V/cm; 17 msec, 0°C, 5 min preincub.,10 min postincub., PBS	Up to 300 transf./10⁶ cells, linear > supercoiled, low temperature favorable, few copy number (1-15) integrated, mitotic arrest by colcemid favorable

Ref.	Year	Authors	Recipient cells	Plasmid /gene	Pulsing CD:E_0;τ	Results
105	1985	Prochownik	Mouse M12 myeloma cells (transient expression)	Plasmid carrying CAT gene	Pulse: ISCO power supply (acc. To Potter) 0°C. PBS, Cuvette	Successfully transformation, CAT actively after 48 hr observed
106	1985	Zerbib et al.	Hamster CHO tk⁻ cells in suspension or monolayer	pAGO with tk gene from herpes virus	CD: 3×6 KV/cm; τ= 10 μsec (20°C, HBS) Square pulse: 3×1.5 KV/cm; 50 μsec(low ionic strength)	150 transf./10^6 cells/μg DNA, threshold: >4 KV/cm 70 transf./10^6 cells/μg DNA,4 plasmids/transformed cell in monolayer
107	1986	Weir and Leder	Mouse B and pre-B cell lines	Functionally rearranged VκII gene	Pulse: ISCO power supply(acc.to potter)	Gene successfully introduced both transiently and permanently by electroporation
108	1986	Yancopoulos et al.	Tk⁻ derivative of 38B9 A-MuLV-transformed pre-B cell line	T cell receptor variable region gene segments on special plasmid construct	Pulse: ISCO power supply (acc. To potter) 0°C, PBS	Linearized plasmid successfully transfected
109	1987	Boston et al.	Daucus carota protoplasts (W001C)	pCATTi, pCATZ2(supercoiled)	Pulse: ISCO power supply(acc. To Potter) Preincub. 5 min 45°C + 5 min on ice with PEG; postincub. 10 min at RT; PCM; Cuvette with AL foil electrodes (acc. To Potter)	2.0 KV setting results in 40% intact viable cells and maximum CAT activity; presence of PEG is necessary (no sharp optimum related to concentration); no effect of heat-shock treatment; linear DNA and presence of carrier DNA decreases CAT expression
110	1992	Puchalski et al.	COS-M6 Monkey kidney cells	Glutathione Stransferase (GST) gene.	(University of Wisconsin Medical Electronics, Madison, WI) (4 0C, 1-cm-wide aluminum electrodes, and 1-cm gap)	With lipofection, only 1% of the surviving cells expressed recombinant GST, although 2.540% of the cells that survived transfection formed colonies.
111	1996	Heller et al.	Rat liver tissue	Psv-β-galactosidase. The BamHI-XhoI fragment carrying the Luc coding sequence from pGEM- Luc was	DC generator, (T820,BTX,Inc.;San Diego, CA) and a switch box (195-7460; BTX,Inc.;San Diego, CA). Field strength 1000V/cm, 6 pulses, duration=99 μs	Gene transfer by electroporation in Vivo may avoid anatomical constraints and low transfection efficiency.

Ref.	Year	Authors	Recipient cells	Plasmid /gene	Pulsing CD:E_0;τ	Results
				cloned into pRc/ CMV plasmid		
112	2000	Dujardin et al.	Rat keratinocytes	pEGFP-N1 with CMV promoter	Cytopulse PA-4000 (Cyto Pulse Sciences, Inc., Maryland, USA), 10 pulses of 1000V and 100μs duration	A localized expression of GFP was observed for at least 7 days in the epidermis. Skin viability was not compromised by electroporation
113	2004	Yamauchi et al.	Human embryonic kidney cells, HEK293	pEGFP-C1 and pDsRed-C1	ElectroSquare-Porator T820, BTX, San Diego, 100v/cm, 10ms	Efficient to transfer multiple genes, in parallel, into cultured mammalian cells for high-throughput reverse genetics research.
114	2006	Yamaoka et al.	Male Japanese white rabbits (2.5–3.0 kg body wt; Kyudo, Tosu, Saga, Japan)	Plasmid DNA	Electric pulse generator (model CUY 201 BTX) P_{on}=5ms, P_{off}=95 ms, No of pulse 10	Optimal gene transfer efficiency in the in situ jugular veins of rabbits, and transgene expression was observed primarily in endothelial cells.
115	2008	Takei et al.	MKN-1, PC-3, F12	VEGF Si RNA	Square Electro Porator (CUY21; Nepagene).	The delivery efficiency correlated to the electric current. The electric current correlated to the microvascular density and vascular endothelial growth factor (VEGF) expression and exhibited a threshold that guaranteed efficient delivery.
116	2010	Kaufman et al.	A549 cells (ATCC, Manassas, VA,USA) a human lung adenocarcinoma cell line	Plasmid DNA	BTX ECM 830 , Electroporation coupled with a Petri-Pulser PP35–2P electrode (Harvard Apparatus, Holliston, MA, USA) using a single 10 ms 160 V square wave	cyclic stretching of the murine lung using ventilation immediately after endotracheal administration and transthoracic electroporation of plasmid DNA increases exogenous gene expression up to fourfold in mice that were not ventilated after plasmid administrationand transfection by electroporation in vivo

Ref.	Year	Authors	Recipient cells	Plasmid /gene	Pulsing CD:E_0;τ	Results
117	2011	Geng et al.	CHO-K1 cells (ATCC)	pEGFP –C1 plasmid(Clontec), cat.no.6084-1	DC power supply (ps350; Standford research system) with alligator chip leads.	Enable to continuous transfection of cells by flow through electroporation in PDMS fluidic channel with alternating wide and narrow section

Table 4. Modified table of gene delivery by electroporation technique. Permission to reprint obtained from Springer book series [94]

In vivo electroporation is a special kind of interest for all researchers because it is nonviral gene delivery with low cost, safety and ease of realization. Recently nucleic acid based gene transfer has been investigated successfully which could be helpful for more clinical trials in human body [118]. T This technique can be applied for food industry [119]. For cancer treatment, electrochemotherapy has emerged and this therapy successfully used for clinical trials [42,99, 120-123]. The different types of application of electroporation has mention below.

5.1.1. Electroporation for DNA transfer

The first reversible electroporation with DNA electrotransfer has been investigated in 1982 [7]. After application of an external electric pulse, cell membrane can permeabilize and DNA will move towards the cell membrane by electrophoretic force and finally it can enter into cyto-plasm of the cell. It has been reported that, small molecules can diffuse into the cell before membrane reseals but DNA cannot transfer inside the cell, if DNA is added immediately after the pulse applications [124]. For better DNA electrotransfer, electric field pulses are important. The electroporating pulse can stimulate a vascular lock (i.e., a transient hypoperfusion) as well as affects the blood circulation to the electropulsed tissues, caused by histamine dependent physiological reaction [125]. For better electrotransfer, electric field pulses have three steps which includes,

(a) Molecules can increase the electrophoretic displacement of the charged molecules due to application of electric filed pulses (b) Cell membrane can enhance the permeabilization (c) Exposed tissues can stimulate the vascular lock [126].

Moreover to deliver the electric pulse for DNA is electrotransfer, just short or high amplitude pulse (e.g. six pulses, 100μs and 1.4 kV cm⁻¹) required to deliver small molecules [127]. For better electrophoretic effect, longer pulses with low amplitude (e.g. eight pulses, 20ms, 200 kV cm⁻¹) are required to increase the transfection rates [124]. However short, high amplitude pulse can follow the long low amplitude pulse. From these two pulses, high amplitude pulse can permeabilize the cell membrane, then long duration low voltage pulse can play the role to drive the DNA into destabilized membrane of the cell [128]. The transfection threshold values are the same for cell electropermeabilization [39]. The transfection efficiency maintains the following equation as mentioned below

$$\text{Transfection Efficiency} = KNT^{2.3}(1 - E_P/E)f(ADN) \qquad (15)$$

where plasmid concentration f(AND) is complex and high level of plasmid is toxic [129] and K is constant. As results, for DNA electrotransfer, the pulse effect (Field strength, short high amplitude pulse, long low amplitude pulse) are very important and which is the major parameters for efficient gene expression into cell and tissues.

5.1.2. Electrotransfer for clinical developments

The electroporation technique has been used widely for transfection of plasmid in vitro and in vivo. Recently this technique has been used for application of DNA vaccine and gene therapies for clinical trials. Electroporation technology are not only the basis for human studies, but also it influence veterinary medical for animals, which can make the bridge between human and animal studies [130-134]. In this section, different clinical trials with electroporation techniques are mentiond below.

5.1.2.1. DNA vaccine

DNA vaccines have excellent potential as preventive or therapeutic agents against cancers and infectious diseases. For a successful DNA delivery into the cell or tissues, DNA must need to subsequently achieve gene expression of the encoded protein at desired level or for the desired duration of time. In vivo electroporation, which can enhance the delivery efficiency and the cellular uptake of an agent by 1,000 times and it can increase the levels of gene expression (i.e. production of the coded protein) by 100 times or more compared to plasmid DNA delivered without other delivery enhancements. DNA vaccination by electroporation technique has been developed in last several years [134-140]. For DNA vaccination by electroporation, preclinical trials for mouse studies revealed that xenogeneic DNA vaccination with gene encoding tyrosinase family membrane can induced antibody and cytotoxic T cell responses resulted in tumor rejection [141-142]. DNA vaccine, p.DOM-PSMA encoded a domain (DOM) of fragment C of tetanus toxin to induced CD[4+] T cell helps to fuse to a tumor-derived epitope from prostate-specific membrane antigen (PSMA) for use in HLA-A2[+] patients with recurrent prostate cancer [139]. For this open level phase I/II work, DNA was delivered by intracellular injection followed by electroporation with five patients per dose level. Plasmid DNA vaccination using electroporation able to elicited robust humoral and CD[8+] T-cell immune responses, while limited invasiveness of delivery [140]. DNA delivered method which included phase I clinical trial investigated safety and immunogenicity of xenogenic tyrosinase DNA vaccine, administered intramuscularly with electroporation to patient with stage IIB, IIC,III or IV melanoma(Clinical Trials. Gov ID NCT00471133). Electroporation with xenogeneic tyrosinase DNA vaccine can increase the human response and anti-tumor effects compared to the vaccine alone [143].

5.1.2.2. RNA electroporation

The RNA transfer by electroporation technique has been increases continuously recently. RNA can enter inside the cell alone or be used for transfection of dendritic cells, showing several advantages as a vaccine including feasibility, applicability, safeness, and effectiveness, when it comes to the generation of immune responses. In vitro experiment, dendritic cells (DC) pulsed with whole tumor RNA or RNA encoding specific antigen like TAAs induced the generation of specific positive cytotoxic T lymphocytes (CTLs) into the cell [144]. Electroporated monocyte-derived DCs with whole RNA from LP-1, U266 cell lines and induced specific CTLs that lysed LP-1 and U266 myeloma cells [145]. The RNA delivery into the dendritic cells (DCs) can be achieved by using electroporation of dendritic cells in the presence of RNA [146]. In vivo study was performed by TriMix dendritic cells (DCs) [146]. The transfection of Dendritic cells (DCs) were performed by electroporation technique with mRNA encoding CD40L, CD70, and a constitutively actived TLR4 as enhancing elements. Additionally the cells were electroporated with either Mage-A3, Mage-C2, tyrosinase, or gp100 mRNA. The intradermal injections at four different sites of 1.25×10^7 TriMix DCs per antigen were provided to two melanoma patient in four times per week. While antigen-specific CD8 T cell responses was detected in both patients after finished treatment, but no data published for this trials [147].

5.1.2.3. HIV vaccine

Since early 1980s, for causative agent of acquired immudeficiency syndrome (AIDS), an effective vaccine has been continuously tried to find to recover AIDS. Nowadays, the HIV vaccine is introduced by electroporation technique. In vivo experiment on mice, electroporation technique can amplify cellular and humoral immune response to a HIV type 1 EnvDNA vaccine, capable of tenfold reduction in vaccine dose and resulting in an increased recruitment of inflammatory cells [148]. The plasmid HIV vaccine, ADVAX env/gag+ADVAX pol/nef-tat (ADVAX), ongoing to examine in phase I trials for uninfected adults (Clinical Trials.gov Identifier: NCT00249106) combination with electroporation as a potential protective vaccine against HIV (Clinical Trials.gov Identifier:NCT00545987). Now more recent study was going on for safety and immunogenicity of an IM injection of two dose of ADVAX using Electroporation TriGrid Delivery Systems (Inchor Medical Systems, Clinical Tials.gov Identifier: NCT00545987)[143].

5.1.2.4. Cancer treatment

Electroporation technique for cancertreatment (Electrochemotherapy) have been increasing rapidly after first reported of clinical use of electroporation [122]. Electrochemotherapy can combine electroporation and chemotherapeutic agents [149-150]. The treatment of cutaneous and subcutaneous tumors has reached for clinical trials using bleomycin or cisplatin by antitumor electrochemotherapy process [120,151-157]. For localized therapy to avoid systematic drug delivery, bleomycin can be injected directly into the tumors by using electrochemotherapy process. Bleomycin is hydrophilic in nature, which can be internalized in limited amounts only in normal condition [158]. The use of bleomycin for electroporation process can directly enter into cytosol and its cytotoxicity can be increased up to

300-5000 fold [159-161]. Different types of cancer can be treated by electroporation technique. The prostate cancer is one of the most common cancer, which is increasing day to day. For this cancer prostate specific antigen (PSA), targeted to the prostate cancer cell for immunotherapeutic approach. The phase I clinical trials with PSA DNA vaccine for human prostate cancer is safe and which can include cellular and humoral immune responses against PSA protein [162-163]. The PSA-DNA vaccine has been investigated by electroporation technique [164-165]. Electroporation treated with CD^{4+}, CD^{8+} cells and antibodies were detected in patient successfully with safe and tolerated mode. Electrochemotherapy has also been investigated for treatment of human colorectal cell line and liver tumours [166-167]. The local treatment of electrochemotherapy (ECT) with master cell tumours of Dog has been experimented in where size of the tumors was 5.2 cm^3 and 2.9 cm^3 treated by surgery and ECT. The ECT treatment was easy, effective and safe local treatment for master cell tumors of Dogs [168]. Recently, electrochemotherapy has been developed in more advancement for treat ment of internal tumors using surgical procedures, endoscopic routes or percutaneous approaches to gain access to the treatment area [169-170].

5.1.3. Skin electroporation

Molecule or DNA vaccine can transport across targeted tissue of human skin is of great interest for transdermal drug delivery and non-invasive chemical sensing. Skin has capability to produce therapeutic molecules, which not only acts as a systematically or locally, but it can create immunological response, when antigen presenting cells will be targeted. The skin containing antigen presenting cells like dendritic cells, langerhans cells, and mononuclear cells. The gene delivery through the skin electroporation is feasible, efficient and comparable to other tissues [171]. The first skin electroporation study was observed in Newborn mice which transfected with a plasmid coding for a neomycin resistance gene [172]. The transfection efficiency can depend upon the age of the skin, where the higher transfection efficiency can be achieve for younger mice compare to the older mice [173]. Skin electroporation, only clinical study has been reported belonging to metastatic melanoma [118]. To date, the skin electroporation has been studied broadly for animal infectious diseases. For most cases Hapatitis B has been investigated for animals through skin electroporation [137,174-176]. Also experiments have been performed vaccine against HIV, smallpox, malaria [177-179].

5.2. Single cell electroporation

By using single cell electroporation technique, it is possible to deliver the molecules such as drugs, DNA, RNA, peptide, nucleic acid into the cell membrane in vivo and vitro for single cell analysis. The plasmid delivery inside the cell membrane with high efficiency in adherent cells and tissues has been studied in vitro [180-184] and in vivo [52,183-186]. Fig.7. show the different applications of single cell electroporation, where membrane can permeabilized to transport protein, small and large molecules inside the single cell.

When two single cells are closed to each other, then cell fusion can occur. Due to high electric field strength, which exceeds the critical value of cell membrane, irreversible electroporation can occur, resulting in cell membrane rapture and finally cell death. This electroporation

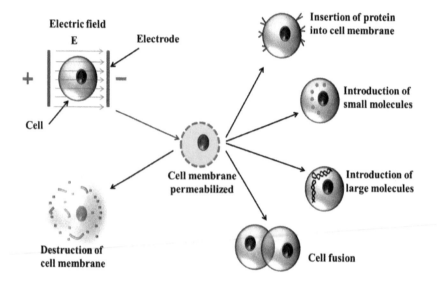

Figure 7. Different application of single cell electroporation. When external applied electric field reaches to the threshold values of the cell membrane, then cell membrane can permeabilized to deliver protein, small and large molecules inside the cell. If two single cells are close to each other, then cell fusion can occur. To apply an intense electric field, which exceeds certain critical value, irreversible electroporation can occur resulting cell membrane rapture and finally cell death. Figure has redrawn from reference. Figure has redrawn with reprint permission obtained from Springer [187].

successfully investigated cell to cell intracellular biochemical variation from millions of cells. However this technique needs a lot of research in the future for more improvement because this technology is in underdeveloped stage. For intracellular targeting, single cell electroporation based systems can be developed for genomic characterization, where a tagged antisense oligonucleotide is introduced to block expression and proteins can be profile by tagged markers [188]. To reduce the electrode size in nanoscale label, selective manipulation of single organelles within a cell can be possible. Thus the localized single cell membrane electroporation (LSCMEP) concept has come in frontier research in last several years [61-63]. This technique can control spatial-temporal process successfully and its have ability to monitor the transfection results in real time situation. To reduce the electrode size in nanoscale label, effective electroporation region should be reduce. As results transfection efficiency should be increase with high cell viability. Florescent markers with single cell electroporation permits direct visualization of cell morphology, cell growth, and intracellular events over timescales ranging from seconds to days. Fluorescent dye or plasmid DNA can enter the neurons with the intact brain of albino Xenopus tadpoles [189]. Individual neurons can be elctroporated by this technique in vivo and in vitro including mature and fully differentiated neurons. The transfection of neurons into brain slices and in intact brains of living animals is possible to use this technique. The neuron transfection achievable up to 1 mm dip into a tissue and electro-physiological recording of individual neuron was possible by use of SCEP [190]

6. Conclusions

This chapter described the detailed concepts about bulk electroporation(BEP) as well as single cell electroporation (SCEP) techniques. In both electroporation technique different types of exogenous molecules such as DNA, RNA, proteins, anticancer drugs, ions, oligonucleotides can be transported into the cell cytosol in vivo or in vitro. For bulk electroporation, the clinical development of DNA based vaccine and immunotherapeutic delivery is progressing. As a nonviral gene transfer, this technique is important for clinical gene transfer regarding efficacy and safety issue compared to other gene transfer techniques. The new technique such as single cell electroporation (SCEP) makes the possibility to judge cell to cell variations with their organelles and intracellular biochemical effect. The development of SCEP technique at clinical level and for biomedical application needs more research in the future. In SCEP, there still lacks the are lack of understanding of theory and molecular delivery inside the cell. But this technique can initiate new root of research, such as single cell biophysics and drug delivery inside single cell. To reduce the electrode gap at nanoscale level, it is possible to do localized single cell membrane electroporation (LSCMEP) by which selective specific single cell organelles can be manipulated with higher transfection rate and high cell viability.

Acknowledgements

The authors greatly appreciate the financial support from National Science Council (NSC) of Taiwan ROC through National Nanotechnology and Nanoscience Program under Contract no. NSC- 98-2120-M-007-003 and NSC 99-2120-M-007-009.

Author details

Tuhin Subhra Santra[1], Pen-Cheng Wang[2] and Fang Gang Tseng[1,2,3]

1 Institute of NanoEngineering and Microsystems (NEMS), National Tsing Hua University, Hsinchu, Taiwan

2 Department of Engineering and System Science (ESS), National Tsing Hua University, Hsinchu, Taiwan

3 Division of Mechanics, Research Center for Applied Sciences, Academia Sinica, Taipei, Taiwan

References

[1] U.Zimmermann, G.Pilwat, F.Friemann, Dielectric breakdown of cell membranes, Bio-phys.J. 1974;14: 881-899.

[2] H.A.Pohl, J.S.Crane, Dielectrophoresis of cells, Biophys.J 1971;11: 711-727.

[3] A.J.H.Sale, W.A.Hamilton, Effect of high electric fields onmicroorganisms II. lysis of erythrocytes and protoplasts, Biochem. Biophys. Acta 1968;163; 37-43.

[4] J.M.Crowley, Electrical breakdown of bimolecular lipid membranes as an electrome-chanical instability, Biophys.J. 1973;13: 711-714.

[5] H.G.L.Coster, A quantitative analysis of the voltage-current relationship of fixed charge membranes and the associated property of "punch throygh", Biophys.J. 1965;5: 669-586.

[6] A.J.H. Sale, W.A.Hamilton, Effects of high electric fields on microorganisms I. killing of bacteria and yeasts, Biochem. Biophys. Acta 1967;148: 781-788.

[7] E.Neumann, M.S-Ridder, Y.Wang, P.H.Hofschneider, Gene transfer into mouse lyo-ma cells by electroporation in high electric fields. EMBO.J. 1982;1: 841-845.

[8] J.C.Weaver, Electroporation of cell and tissues, IEEE transactions on plasma science 2000; 28(1): 24-33.

[9] U.Pliquett, J.C.Weaver, Electroporation of human skin: Simultaneous measurement of changes in the transport of two florescent molecules and in the passive electrical properties, Bioelectrochemistry and Bioenergetics 1996;39(1): 1-12

[10] K.K.Ewert, A.Ahmed, N.F.Bouxsein, H.M.Evans, C.R.Safinya, Non-viral gene deliv-ery with cationic liposome-DNA complexes, Methods Mol. Biol. 2008;433:159-170.

[11] D.H.Fuller, P.Loudon, C.Schmaljohn, Preclinical and clinical progress of particle-mediated DNA vaccines for infectious diseases, Methods 2006;40:86-97

[12] S.Ohta, K.Suzuki, Y.Ogino, S.Miyagawa, A.Murashima, D.Matsumaru, G.Yamada, Gene transduction by sonoporation. Dev Growth Differ 200;, 50(6):517-520.

[13] J.A.O'Brien, S.C.R.Lummis, Biolistic transfection of neuronal cultures using a hand-held gene gun, Nature Protoc. 2006;1:977-981.

[14] M.R.Capecchi, Highefficiency transformation by direct microinjection of DNA into cultured mammalian cells, Cell 1980; 22: 479-488.

[15] A.Naga, M.Gertsenstein, K.Vintersten, R.Behringer, Manipulating the Mouse Em-bryo: A Laboratory Manual(Cold Spring Laboratory, 2003)

[16] D.J.Wells, Gene therapy progress and prospects: electroporation and other physical methods, Gene Ther. 2004; 11:1363-1369.

[17] G.L.Prasanna, T.Panda, Electroporation:basic principles, practical considerations and application in molecular biology, Bioprocess Engineering 1997;16: 261-264.

[18] E.M.Miller, J.A.Nickoloff,Escherichia coli electrotransformation. In: J.A.Nickoloff , editor. electroporation protocols for microorganisms. Totowa, New Jersey: Humana Press. 1995; 105-114.

[19] H.L.Withers, Direct plasmid transfer between bacterial species and electrocuring. In: J.A.Nickoloff, editor. Electroporation protocols for microorganisms. Totowa, New Jersey: Humana Press. 1995; 47-54.

[20] T.Tryfona, M.T.Bustard, Enhancement of biomolecule transport by electroporation: A rivew of theory and practical application to transformation of corynebacterium. Biotechnol.and Bioengg 2006;93(3): 413-423.

[21] J.A.Nickoloff (edited), Methods in molecular Biology, Animal cell electroporation and electrofusion protocols, Humana press Inc. Totowa, NJ, Vol 48(1995) ISBN-10 / ASIN: 089603304X ,ISBN-13 / EAN: 9780896033047.

[22] J.Teissie, N.Eynard, B.Gabriel, M.Rols, Electropermeabilization of cell membranes. Adv Drug Deliv Rev 1999;35: 3-19.

[23] M.Pavlin, N.Pavselj, D.Miklavcic, Dependence of induced transmembrane potentialon cell density, arrangements, and cell position inside the cell system. IEEE Trans Biomed Eng 2002;49: 605-612.

[24] E.T.Jordan, M.Collins, J.Terefe, L.Ugozzoli, T.Rubio, J. Biomolecular Techniques 2008;19: 328-334.

[25] J.A.Sunders, C.H.Lin, B.H.Hou,J.Cheng, N.Tsengwa,J. JLin, C.R.Smith, M.S.McIntosh, S.V.Wert, Molecular biotechnology 1995;3: 181-190.

[26] K.Vafai (Editor),Porous media, Applications in biological systems and biotechnology, CRC press, Taylor and Francis group, Boca Raton, London, Yew York.2011

[27] R.Lanza, R.Langer, J.P.Vacanti, Principle of tissue engineering, second edition, academic press,London UK,2002.

[28] O.C.Tuncay, D.Ho, M.K.Barker, Oxygentension regulates osteoblast function. American Journal of orthodontics and dentofacial orthopedics 1994;105: 457-463.

[29] T.Arnett, D.Gibbons, J.Utting, I.Orriss, A.Hoebertz, M.Rosendaal, S.Mejhji, Hypoxia is a major stimulator of osteoclast formation and bone resorption, Jurnal of Cellular Physiology 2003;196: 2-8.

[30] M.Muraji, W.Tatebe, H.Berg, The influence of extracellular alkali and alkaline earth ions on electropermeabilization of saccharomyces cerevisiae. Bioelectrochem Bioenergetics 1998;46: 293-295.

[31] W.Y.Gu, W.M.Lai, V.C.Mow,A mixture theory for charged hydrated soft tissues containing multi-electrolytes: passive transport and swelling behavior, Journal of Biomechanical Engineering 1998;120: 169-180.

[32] A.Maton, J.Hopkins, S.Johnson, D.LaHart, M.Q.Warner, J.D.Wright, Cells buildings block of the life, Prentice Hall, Upper Saddle River, NJ, 1997.

[33] J.M.Huyghe, J.D.Janssen, Quadriphasic mechanics of sweeling incompressible porous media, International Journal of Engineering Science 1997;35: 793-802.

[34] T.Ohshima, M.Sato, Bacterial sterilization and intracellular protein release by pulsed electric field. Adv. Biochem Eng Biotechnol 2004;90: 113-133.

[35] [35]K.Kinosita, T.Y.Tsong, Formation and resealing of pores of controlled sizes in human erythrocyte membrane, Nature 1977b;286: 438-441.

[36] B.J.Wards, D.M.Collins, Electroporation at elevated temparatures substantially improved transformation efficiency of slow growing mycobacteria, FEMS Microbiol Lett 1996;145: 101-105.

[37] M.P.Rols, C.Delteil, G.Serin,, J.Teissie, Temparature effects on electrotransfection of mammalian cells, Nucleic Acids Res 1994;22: 540.

[38] S.Fielder, R.Wirth, Transformation of bacteria with plasmid DNA by electroporation, Analyt Biochem 1988;170: 38-44.

[39] H.Wolf, M.P.Rols, EBoldt, E.Neumann, J.Teissie, Control by pulse parameters of electric field-mediated gene transfer in mammalian cells, Biophys J 1994;66(2): 524-531.

[40] D.J.Winterbourne, S.Thomas, J.Hermon-Taylor, I.Hussain, A.P.Johnstone, Electric shock-mediated transfection of cells. Characterization and optimization of electrical parameters, Biochem J 1988;251: 135-160.

[41] E.Neumann, S.Kakorin, I.Tsoneva, B.Nikolova, T.Tomov, Calcium –mediated DNA adsorption to yeast cells and kinetics of cell transformation by electroporation, Biophys J 1996;71: 868-877.

[42] J.Gehl, Electroporation: theory and methods, perspectives for drug delivery, gene therapy and research, Acta Physiol Scand 2003;177: 437-447.

[43] E.Neumann, B.Rosenheck, Permibility induced by electric impulsions in vesicular membranes, J.Membr.Biol 1972;10; 279-290.

[44] K.H.Schoenbach, S.J.Beebe, E.S.Buescher, Intracellular effect of ultrashort electrical pulses, Bioelectromagnetics 2001;22: 440-448.

[45] B.Gabriel, J.Teissie, Direct observation in the millisecond time range of fluorescent molecule asymmetrical interaction with the electropermeabilized cell membrane, Biophys J 1997;73: 2630-2637.

[46] J. A. Lundqvist, F. Sahlin, M. A. Aberg, A. Strimberg,P. S. Eriksson and O. Orwar, Altering the biochemical state of individual cultured cells and organelles with ultra-microelectrodes, Proc. Natl. Acad. Sci 1998; 95: 10356–10360.

[47] E.Takle, R.D.Astumian, P.B.Chock, Electro-permeabilization of cell membranes: effect of the resting membrane potential, Biochem Biophys Res Commun 1990;172: 282-287.

[48] F.J.Chang, T.S.Reese, Change in membrane structure induced by electroporation as revealed by rapid-freezing electron microscopy, Biophys J 1990;58: 1-12.

[49] S.Y.Ho, G.S.Mittal, Electroporation of cell membrane: A review. Crit Rev Biotechnol 1996;16(4): 349-362.

[50] J.Teissie, M.Golzio, M.P.Rols, Mechanisms of cell membrane electropermeabilization: A minireview of our present (lack of ?) knowledge, Biochimica et Biophysica Acta 2005;1724: 270-280.

[51] D.Ovcharenko, R.Jarvis, S.Hunicke-Smith, K.Kelnar, D.Brown, High throughput RNAi screening in vitro: From cell lines to primary cells, RNA 2005; 11; 985-993.

[52] J.E.Bestman, R.C.Ewald, S-L Chiu, H.T.Cline, In vivo single cell electroporation for transfer of DNA and macromolecules, Nature Protocols 2006;1(3): 1268-1272.

[53] M.Khine, A.Lau, C.I-Zanetti, J.Seo, L.P.Lee, Asingle cell electroporation chip, Lab on a chip 2005;5: 38-43.

[54] D.C.Chang, B.M.Chassy, J.A.Saunders, A.E.Sowers, Guide to Electroporation and Electrofusion, Academic Press, Inc.,San Diego, 1992.

[55] M.Wang, O.Orwar, J.Olofsson,S.G.Weaber, Single cell electroporation, Anal Bioanal Chem 2010;397: 3235-3248.

[56] F.Ryttsen, C.Farre, C.Brennan, SS.G.Weber, K.Nolkrantz, K.Jardmark, D.T.Chiu, O.Orwar, Characterization of single cell electroporation by using Patch-Clamp and Flouorosence microscopy, Biophysical Journal 2000;79: 1993-2001.

[57] K.Nolkrantz, C.Farre, A.Brederlau, R.I.D.Karlsson, C.Brennan, P.S.Erriksson, S.G.Weber, M.Sandberg, O.Orwar, Electroporation of single cells and tissues with an electrolyte-filled capillary, Anal Chem 2001,73: 4469-4477.

[58] K. Haas, W. C. Sin, A. Javaherian, Z. Li and H. T. Cline, Single cell electroporation for gene transfer in Vivo, Neuron 2001;29: 583–591.

[59] Y. Huang and B. Rubinsky, Microfabricated electroporation chip for single cell membrane permeabilization, Sens. Actuators A 2001;89: 242–249.

[60] K. Nolkrantz, C. Farre, K. J. Hurtig, P. Rylander and O. Orwar, Functional screening of intracellular proteins in single cells and in patterned cell arrays using electroporation, Anal. Chem 2002;74: 4300–4305.

[61] D.Nawarathna, K.Unal, H.K.Wickramasinghe, Localized electroporation and molecular delivery into single living cells by atomic force microscopy,Appl. Phys. Lett 2008;93: 15311-15313.

[62] P.E.Boukany, A.Morss, W-C Liao, B.Henslee, H.Jung, X.Zhang, B.Yu, X.Wang, Y.Wu, L.Li, K.Gao, X.Hu, X.Zhao, O.Hemminger, W.Lu, G.P.Lafyatis, L.J.Lee, Nature Nanotechnology 2011;6: 747-754.

[63] S-C. Chen, T.S.Santra, C-J Chang, T-J Chen, P-C Wang, F-G Tseng, Biomedical Microdevices, DOI 10.1007/s10544-012-9660-9.

[64] S.J.Singer, G.L.Nicolson, The fluid mosaic model of the structure of cell membranes, Science 1972;175: 720-731.

[65] I.G.Abidor, V.B.Arakelyan, L.V.Chernomordik, Y.A.Chizmadzhev, V.F.Pastushenko, M.R.Tarasevich, Electric breakdown of bilayer lipid membrane.I. The main experimental facts and their qualitative discussions, Bioelectrochem.Bioenerg 1979;6: 37-52.

[66] J.C.Weaver,R.A.Mintzer, Decreased bilayer stability due to transmembrane potentials, Phys. Lett 1981;86A: 57-59.

[67] I.P.Sugar, The effects of external fields on the structure of lipid bilayers, J.Physiol. Paris 1981;77: 103-1042.

[68] K.T.Powell, J.C.Weaver, Transient aqueous pores in bilayer membranes: A statistical theory, Bioelectrochem. Bioenerg 1986;15: 211-227.

[69] J.D.Litster, Stability of lipid bilayers and red blood cell membranes, Phy. Lett 1975;53A: 193-194.

[70] A.G.Petrov, M.D.Mitov, A.I.Derzhanski,, Edge energy and pore stability in bilayer lipid membranes , in :Advances in Liquid crystal research and applications (L. Bata, ed.), Pergamon Press, Elmsford, NY 1980;695-737.

[71] L.V.Chernomordik, S.I.Sukharev, I.G.Abidor, Y.A.Chizmadzhev, Breakdown of lipid bilayer membranes in an electric field, Biochim. Biophys. Acta 1983;736: 202-213.

[72] J.Panyam, V.Labhasetwar, Dyanamics of endocytosis and exocytosis of poly (D,L-Lactide-co-Glycolide) nanoparticles in vascular smoothmuscle cells. Pharmaceutical Research 2003;20(2): 212-219.

[73] I.R.Miller, Structural and energetic aspects of change transport in lipid bylayers and biological membranes, in: Topics in Bioelectrochemistery and Bioenergetics(G. Milazzo, ed.), Wiley, New York 1981; 161-224.

[74] J.Gimsa, D.Wachner, Analytical description of the transmembrane voltage induced on arbitrarily oriented ellipsoidal and cylindrical cells *Biophysical J* 2001;81: 1888-1896.

[75] T.Kotnik, D.Miklavcic, Analytical description of transmembrane voltage induced by electric fields on spheroidal cells, Biophysics. J 2000;79: 670-679.

[76] C.Polk, E.Postow, Eds., CRC Handbook of biological effects of electromagnetic fields, 2nd ed. Boca Ratom, FL:CRC, 1996.

[77] E.A.Gift, J.C.Weaver, Observation of extreamly heterogeneouselectroporative uptakewhich changes with electric field pulse amplitude in saccharomyces cerevisiae, Biochem. Biophys.Acta 1995;1234: 52-56.

[78] J.C.Weaver, Yu.A.Chizmadzhev, Theory of electroporation: A review, Bioelectrochem. Bioenerget 1996;41: 135-160.

[79] H.T.Tien, Bilayer lipid membrane (BLM), Dekker, New York 1974.

[80] R.Benz, F.Beckers, U.Zimmermann, Reversibale electrical breakdown of lipid bilayer membranes: A charge pulse relaxation study, J.Membr.Biol 1979;48: 181-204.

[81] R.Benz, U.Zimmermann, Relaxation studies on cell membranes and lipid bilayers in the high electric field range, Bioelectrochem.Bioenerg 1980b;7: 723-739.

[82] T.Y.Tsong, T.T.Tsong, E.Kingsley, R.Siliciano, Relaxation phenomena in human erythrocyte suspensions, Biophys J 1976;16: 1091-1104.

[83] P.Linder, E.Neumann, K.Rosenheck, Kineticsof permeability changes induced by electric impulses in chromaffin granules J.Membr.Biol 1977;32: 231-254.

[84] K.S.Cole, Membranes, Ions and Impulses, University of California Press, Berkeley, 1972

[85] H.Pauly, W.P.Schwan, Dielectric properties and ion mobility in erythrocytes, Biophys.J 1966;6: 621-639.

[86] R.Stampfli, Reversable electrical breakdown of the excitable membrane of a Ranvier node, Ann. Acad. Bras. Cien 1958;30: 57-63.

[87] U.Zimmermann, Electric field-mediated fusion and related electrical phenomena, Biochim.Biophys. Acta 1982;694: 222-227.

[88] L.V.Chernomordik, S.I.Sukharev, S.V.Popov, V.F.Pastushenko, A.V.Sokirko, I.G.Abidor, Y.A.Chizmadzhev, The electrical breakdown of cell and lipid membranes: The similarity of phenomenologies, Biochim, Biophys. Acta 1987;902: 360-373.

[89] R.Benz,U.Zimmermann, The resealing process of lipid bilars after reversible electrical breakdown, Biochim. Biophys. Acta 1981;640: 169-178.

[90] I.G.Abidor, S.I.Sukharev, L.V.Chernomordik, Y.A.Chizmadzhev, The reversible electrical breakdown of bilayer lipid membranes modified by UO_2^{2+} ions, Bioelectrochem.Bioenerg 1982;9: 141-148.

[91] L.V.Chernomordik, S.I.Sukharev, I.G.Abidor, Y.A.Chizmadzhev, The study of the BLM reversible electrical breakdown mechanism in the presence of UO_2^{2+} ions, Bioelectrochem.Bioenerg 1982;9: 149-155.

[92] [92]S.I.Sukharev,L.V.Chernomordik, I.G.Abidor, Reversable electrical breakdown of holoturinmodified bilayer lipid membranes, Biofizika 1983;28: 423-426.

[93] A.Barnett, J.C.Weaver, Electroporation:a unified, quantitative theory of reversible electrical breakdown and repture, Bioelectrochem. Bioenerg 1991;25: 163-182.

[94] A.E.Sowers, C.A.Jordan, "Electroporation and elecrofusion in cell biology" Edited by E.Neumann, Planam press, New York 1989; 308-315.

[95] [95]B.Rubinsky, Irreversible electroporation in medicine, Technol. In Cancer Research Treatment, ISSN 1533-0346, 2007;6: 255-259.

[96] J.A.Nollet, Researches sur les causes particulieres des phenomenes electriques, Paris: Chez H.L. Guerin &L.F.Delatour 1754.

[97] R.W.Glaser, A.Wagner, E.Donath, Volume and ionic composition changes in erythrocytes after electric breakdown: Simulation and experiment, Bioelectrochem. Bioenerg 1986;16: 455-470.

[98] L.V.Chernomordik, M.M.Kozlov, G.B.Melikyan, I.G.Adibor, V.S.Markin, A.Y.Chizmadzhev, The shape of lipid molecules and monolayer membrane fusion, Biochim, Biophys. Acta 1985;812: 643-655.

[99] A.Gothelf, L.M.Mir, J.Gehl, Electrochemotherapy: Results of cancer treatment using enhanced delivery of bleomycin by electroporation, Cancer Treat. Rev 2003;29: 371-387.

[100] S.Orlowski, L.M.Mir, Cell electropermeabilization: a new tool for biochemical and pharmacological studies, Biochim. Biophys. Acta 1993;1154: 51-63.

[101] M.Golzio, M.P.Rols, J.Teissie, In vitro and in vivo electric field-mediated permeabilization, gene transfer, and expression, Methods 2004;33: 126-135.

[102] N.Shivarova, W.Forster, H-E.Jacob, R.Grigorova, Microbiological implications of electric field effects. VII. Stimulation of plasmid transformation of Bacillus cereus protoplasts by electric field pulses, Z.Allg. Mikrobial 1983;23: 595-599.

[103] F.G.Falkner, E.Neumann, H.G.Zachau, Tissue specificity of the initiation of immunoglobulin κ gene transcription, Hoppe-Seyler's Z physiol. Chem 1984;365: 1331-1343.

[104] H.Potter, L.Weir, P.Leder, Enhancer-dependent expression of human κ immunoglobulin genesintroduced into mouse pre-B lymphocytes by electroporation, Proc. Natl. Acad. Sci. USA 1984;81: 7161-7165

[105] E.V.Prochownik, Relationship between an enhancer element in the human antithrombin III gene and an immunoglobulin light-chain gene enhancer, Nature 1985;316: 845-848.

[106] D.Zerbib, F.Amalric, J.Teissie, Electric field-mediated transformation: Isolation and characterization of a TK$^+$ subclone, Biochem. Biophys. Res. Commun 1985;129: 611-618.

[107] L.Weir, P.Leder, Structure and expression of a human subgroup II immunoglobulin κ gene,Nucleic Acids Res 1986;14: 3957-3970.

[108] G.D.Yancopoulos, T.K.Blackwell, H.Suh, L.Hood, F.W.Alt, Introduced T cell receptor variable region gene segments recombine in pre-B cells: Evidence that B and Tcells use a common recombinase, Cell 1986;44: 251-259.

[109] R.S.Boston, M.R.Becwar, R.D.Ryad, P.B.Goldsbrough, B.A.Larkins, T.K.Hodges, Expression from heterologous promoters in electroporated carrot protoplasts, Plant physiol 1987;83: 742-746.

[110] R.B.Puchalski, W.E.Fahl, Gene transfer by electroporation,lipofection, and DEAE-Dextran transfection: compatibility with cell sorting by flow cytometry, Cytometry 1992;13: 23-30.

[111] R.Heller,M.Jaroszeski, A.Atkin, D.Moradpour, R.Gilbert, J.Wands, C.Nicolau, In vivo gene electroinjection and expression in rat liver, FEBS Letters 1996;389: 225-228.

[112] N.Dujardin, P.V.D.Smissen, V.Preat, Topical gene transfer into rat skin using electroporation, Pharmaceutical Research 2000;18(1): 61-66.

[113] F.Yamauchi, K.Kato, H.Iwata, Spetially ad temporally controlled gene transfer by electroporation into adherent cells on plasmid DNA –loaded electrodes, Nucleic Acids Research 2004;32(22): e187

[114] T.Yamaoka, Y.Yonemitsu, K.Komori, H.Baba, T.Matsumoto, T.Onohara, Y.Maehara, Ex vivo electroporation as a potent new strategy for nonviral gene transfer into autologous vein grafts, Am J Physiol Heart Circ Physiol 2005;289: H1865–H1872.

[115] Y.Takei, T.Nemoto, P.Mu, T.Fujishima, T.Ishimoto, Y.Hayakawa, Y.Yuzawa, S.Matsuo, T.Muramatsu, K.Kadomatsu, In vivo silencing of a molecular target by short interfering RNA electroporation: tumor vascularization correlates to delivery efficiency, Mol. Cancer Ther 2008;7: 211-221.

[116] C.D.Kaufman, R.C.Geiger, D.A.Dean, Electroporation and mechanical ventilation-mediated gene transfer to the lung, Gene Therpy 2010;17: 1098-1104.

[117] T.Geng, Y.Zhan, J.Wang, C.Lu, Transfection of cells using flow through electroporation based on constant voltage, Nature protocalls 2011;6(8): 1192-1208.

[118] A.I.Daud,R.C.DeConti, S.Andrews, P.Urbas, A.I.Riker, V.K.Sondak,, Phase I trial of interleukin-12 plasmid electroporation in patients with metastatic melanoma, Journal of Clinical Oncology 2008;26: 5896-5903.

[119] J.Teissie, N.Eynard, M.C.Vernhes, A.Benichou, V.Ganeva, B.Galutzov, P.A.Cabanes, Recent biotechnological devlopements of electropulsation.A prospective review, Bio-eletrochemistery 2002;55: 107-112.

[120] L.M.Mir, M.Belehradek, C.Domenge, Electrochemotherapy, a new antitumor treatment:first clinical trial. CR Acad. Sci. III 1991;313: 613-618.

[121] L.M.Mir, S.Orlowski, J.Jr.Belehradek,C.Paoletti, Electrochemotherapy potentiation of antitumour effect of bleomycin by local electric pulses, Eur. J. Cancer 1991;27: 68-72.

[122] M.Belehradek, C.Domenge, B.Luboinski, S.Orlowski,J.Jr.Belehradek,L.M.Mir, Electrochemotherapy, a new antitumor treatment, First clinical phase I-II trial, Cancer 1993;72: 3694-3700.

[123] L.M.Mir, L.F.Glass, J.Teissie, Effetcive treatment of cutaneous and subcutaneous malignant tumours by electrochemotherapy, Br.J.Cancer 1998;77: 2336-2342.

[124] [124]L.M.Mir, M.F.Bureau, J.Gehl, High efficiency gene transfer into skeletal muscle mediated by electric pulses. Proc. Natl Acad Sci. USA 1999;96: 4262-4267.

[125] J.Gehl, T.Skovsgaard, L.M.Mir, Vascular reactions to in vivo electroporation:Characterization and consequences for drug and gene delivery, Biochimica et Biophysica Acta 2002;1569: 51-58.

[126] L.M.Mir, Nucleic acids electrotransfer based gene therapy(electrogenetherapy):past, current and future, Mol Biotechnol 2009;43: 167-176.

[127] R.Haller, M.Jaroszeski, A.Atkin, In vivo gene electroinjection and expression in rat liver, FEBS Lett 1996;389: 225-228.

[128] M.F.Bureau, J.Gehl, V.Deleuze, L.M.Mir, D.Scherman, Importance of association between permeabilization and electrophoretic forces for intramuscular DNA electrotransfer, Biochem Biophys Acta 2000;1474: 353-359.

[129] MP.Rols, D.Coulet, J.Tessie, Highly efficient transfection of mammalian cells by electric field pulses, Application to large volumes of cell culture by using a flow system. Eur. J. Biochem 1992;206: 115-121.

[130] K.Dhama, M.Mahendran, P.K.Gupta, A.Rai,, DNA vaccines and their applications in veterinary practice: current prospectives, Vet Res Commun 2008;32: 341-356.

[131] R.Person, A.M.Bodles-Brakhop, M.A.Pope, P.A.Brown, A.S.Khan, R.Draghia-Akli, Growth hormone-releasing hormone plasmid treatment by electroporation decreases offspring mortality over three pregnancies, Mol Ther 2008;16: 1891-1897.

[132] [132]P.A.Brown,A.Bodles-Brakhop, R.Draghia-Akli, Plasmid growth hormonerelea-singhormone therapy in healthy and laminitis-afflicted horses-evalution and pilot study, J.Gene Med 2008;10: 564-574.

[133] P.A.Brown, A.M.Bodles-Brakhop, R.Draghia-Akli, Effects of plasmid growth hormone releasing hormone treatment during heat streass, DNA Cell Biol 2008;27: 629-635.

[134] A.M.Bodles-Brakhop, P.A.Brown, M.A.Pope, R.Draghia-Akli, Double blinded, placebo-controlled plasmid GHRH trial for cancer-associated anemia in dogs, Mol Ther 2008;16: 862-870.

[135] J.Glasspool-Malone, S.Somiari, J.J.Drabick, R.W.Malone, Efficient nonviral cutaneous transfection, Mol. Ther 2000;2: 140-146.

[136] S.Somiari, J.Glasspool-Malone, J.J.Drabick, Theory and in vivo application of electroporative gene delivery, Mol Ther 2000;2: 178-187.

[137] J.J.Drabick, J.Glasspool-Malone, A.King, R.W.Malone, Cutaneous transfection and immune responses to intradermal nucleic acid vaccination are significantly enhanced by in vivo electropermeabilization, Mol Ther 2001;3: 249-255.

[138] F.Liu, L.Huang, A syringe electrode device for simultaneous injection of DNA and electrotransfer, Mol Ther 2002;5: 323-328.

[139] L.Low, A.Mander, K.McCann, D.Dearnaley, T.Tjelle, I.Mathiesen, F.Stevenson, C.H.Ottensmeier, Human Gene Therapy 2009;20: 1269-1278.

[140] A.Brave, S.Nystrom, A-K.Roos, S.E.Applequist, Plasmid DNA vaccination using skin electroporation promotes polyfunctional CD4 T-cell responces,*Immunology and Cell Biology* 2011;89: 492-496.

[141] L.W.Weber, W.B.Bowne, J.D.Wolchok, R.Srinivasan, J.Qin, Y.Moroi, Tumor immunity and autoimmunity induced by immunization with homologous DNA, J.Clin Invest 1998;102: 1258-1264.

[142] W.B.Browne, R.Srinivasan, J.D.Wolchok, W.G.Hawkins, N.E.Blachere, R.Dyall, Coupling and uncoupling of tumor immunity and autoimmunity, J.Exp Med 1999;190: 1717-1722.

[143] A.M.Bodles-Brakhop, R.Heller, R.D-Akli, Electroporation for the delivery of DNA based vaccines and immunotherapeutics: current clinical devlopments, Molecular Therapy 2009;17: 585-592.

[144] D.Boczkowski, S.K.Nair, D.Snyder, E.Gilboa, Dendritic cells pulsed with RNA are potent antigen-presenting cells in vitro and in vivo, J. Ex. Med 1996;184: 465–472.

[145] C. Milazzo, V. L. Reichardt, M. R. Muller, F. Grunebach, and P. Brossart,Induction of myeloma-specific cytotoxic T cells using dendritic cells transfected with tumor-derived RNA, Blood 2003;101: 977–982.

[146] A. Bonehill, A. M. T. Van Nuffel, J. Corthals, Single-step antigen loading and activation of dendritic cells by mRNA electroporation for the purpose of therapeutic vaccination in melanoma patients, Clin. Can. Res 2009;15: 3366–3375.

[147] A.Bringmann, S.A.Erika Held, A.Heine, P.Brossart, RNA vaccine in cancer treatment, Journal of Biomedicine and Biotechnology 2010;2010: 1-12.

[148] J.Liu, R.Kjeken, I.Mathiesen, D.H.Barouch, Recruitment of antigen-presenting cells to the site of inoculation and augmentation of human immunodeficiency virus type 1 DNA vaccine immunogenicity by in vivo electroporation, J.Virol 2008;82: 5643-5649.

[149] S.B.Dev, Killing cancer cells with a combination of pulsed electric fields and chemotherapeutic agents, Cancer Watch 1994;3: 12-14.

[150] S.B.Dev, G.A.Hofmann, Electrochemotherapy-A novel method of cancer treatment, Cancer Treatment Rev 1994;20: 105-115.

[151] Z.Rudolf, B.Stabuc, M.Cemazar, D.Miklavcic, L.Vodovnik, G.Sersa, Electrochemotherapy with bloemycin: The first clinical experience in malignant melanoma patients, Radiol Oncol 1995;29: 229-235.

[152] [152]V.M.Munoz, G.P.Ortega, Electrochemotherapy for treatment of skin and soft tissue tumours.Update and definition of its role in multimodal therapy. Clin Transl Oncol 2011;13: 18-24 .

[153] T.Hampton, Electric Pulses Help With Chemotherapy, May Open New Paths for Other Agents, JAMA 2011;305: 549-551.

[154] A.Testori,G.Tosti, C.Martinoli, G.Spadola, F.Cataldo, F.Verrecchia, F.Baldini, M.Moscini, J.Soteldo, I.Tedeschi, C.Passoni, C.Pari, A.di. Pietro, P.F.Ferrucci, Electrochemotherapy for cutaneous and subcutaneous tumor lesions: a novel therapeutic approach, Dermatol Ther 2010:23: 651-661.

[155] M.G.Moller, S.Salwa, D.M.Soden, G.C.O'Sullivan, Electrochemotherapy as an adjunct or alternative to other treatments for unresectable or in-transit melanoma. Expert Rev Anticancer Ther 2009;9: 1611-1630.

[156] G.Sersa, D.Miklavcic, M.Cemazar, Z.Rudolf, G.Pucihar, M.Snoj, Electrochemotherapy in treatment of tumours. Eur J Surg Oncol 2008;34: 232-240.

[157] J.O.Larkin, C.G.Collins, S.Aarons, M.Tangney, M.Whelan, S.O'Reily, O.Breathnach, D.M.Soden, G.C.O'Sullivan, Electrochemotherapy - Aspects of preclinical development and early clinical experience. Ann Surg 2007;245: 469-479.

[158] L.M.Mir, O.Tounekti, S.Orlowski, Bleomycin revival of an old drug, Gen Pharmacol 1996;27: 745-748.

[159] S.Orloswski, J.Jr.Belehradek, C.Paoletti, L.M.Mir, Transient electropermeabilization of cells in culture. Increase of the cytotoxicity of anticancer drugs, Biochem Pharmacol 1988;37: 4727-4733.

[160] J.Gehl, T.Skovsgaard, L.M.Mir, Enhancement of cytotoxicity by electropermeabilization: an improved method for screening drugs, Anticancer Drugs 1998;9: 319-325.

[161] M.J.Jaroszeski, V.Dang, C.Pottinger, J.Hickey, R.Gilbert, R.Haller, Toxicity of anticancer agents mediated by electroporation in vitro,Anticancer Drugs 2000a;11: 201-208.

[162] M.Pavlenko, A.K.Roos, A.Lundqvist, A.Palmborg, A.M.Miller, V.Ozenci, A phase I trial of DNA vaccination with a plasmid expressing prostate specific antigen in patients with hormone-refractory prostate cancer, Br. J Cancer 2004;91: 688-694.

[163] A.M.Miller, V.Ozenci, R.Kiessling, P.Pisa, Immune monitoring in a phase 1 trial of a PSA DNA vaccine in patients with hormone-refractory prostate cancer, J.immunother 2005;28: 389-395.

[164] A.Kroos, A.King, P.Pisa, DNA vaccination for prostate cancer, Methods. Mol. Biol 2008;423: 463-472

[165] A.K.Roos, S.Moreno, C.Leder, M.Pavlenko, A.King, P.Pisa, Enhancement of cellular immune response to a prostate cancer DNA vaccine by intradermal electroporation, Mol Ther 2006;13: 320-327.

[166] M.Kambe, D.Arita, H.Kikuchi, T.Funato, F.Tezuka, M.Gamo,Y.Murakawa, R.Kanamaru, Enhancing the effect of anticancer drugs against the colorectal cancer cell line with electroporation, Tohoku, J.Exp.Med 1996;180: 161-171.

[167] M.J.Jaroszeski, R.A.Gilbert, R.Heller, In vivo antitumor effects of electrochemotherapy in a hepatoma model, Biochim. Biophys. Acta 1997;1334: 15-18.

[168] V.Kodre, M.Cemazar, J.Pecar, G.Sersa, A.Cor, N.Tozon, Electrochemotherapy compared to surgery for treatment of canine mast cell tumors, In Vivo 2009;23(1): 55-62.

[169] D.Miklavcic, M.Snoj, A.Zupanic, B.Kos, M.Cemazar, M.Kropivnik, M.Bracko, T.Pecnik, E.Gadzijev, G.Sersa, Towards treatment planning and treatment of deep-seated solid tumors by electrochemotherapy. BioMed Eng OnLine 2010;9: 1-12.

[170] D.M.Soden, J.Olarkin, C.G.Collins, M.Tangney, S.Aarons, J.Piggott, A.Morrissey, C.Dunne, G.C.O'Sullivan, Successful application of targeted electrochemotherapy using novel flexible electrodes and low dose bleomycin to solid tumours. Cancer Lett 2006;232: 300-310.

[171] O.Thanaketpaisarn, M.Nishikawa, F.Yamashita, M.Hishida, Tissue-specific characteristics of in vivo electricgene: transfer by tissue and intravenous injection of plasmid DNA. Pharm Res 2005;22(6): 883–891.

[172] A.V.Titomirov, S.Sukharev,E.Kistanova, In vivo electroporation and stable transformation of skin cells of newbornmice by plasmid DNA. Biochim Biophys Acta 1991;1088(1): 131–134.

[173] S.Chesnoy, L.Huang, Enhanced cutaneous gene delivery following intradermal injection of naked DNA in a highionic strength solution.Mol Ther 2002;5(1): 57–62.

[174] S.Babiuk, M.E.Baca-Estrada, M.Foldvari, L.Baizer, R.Stout,M.Storms, Needle-free topical electroporationimproves gene expression from plasmids administered in porcine skin. Mol Ther 2003;8(6): 992–998.

[175] B.M.Medi, S.Hoselton, R.B.Marepalli, J.Singh, Skin targeted DNA vaccine delivery using electroporation in rabbits. I: efficacy. Int J Pharm 2005;294(1-2): 53–63.

[176] L.Zhang, G.Widera, D.Rabussay, Enhancement of the effectiveness of electroporation-augmented cutaneousDNA vaccination by a particulate adjuvant.Bioelectrochemistry 2004;63(1-2): 369–373.

[177] L.A.Hirao, L.Wu, A.S.Khan, A.Satishchandra, R.Draghia-Akli, D.B.Weiner, Intradermal/subcutaneous immunizationby electroporation improves plasmid vaccine delivery and potency in pigs and rhesus macaques. Vaccine 2008;26(3): 440–448.

[178] J.W.Hooper, J.W.Golden, A.M.Ferro, A.D.King, Smallpox DNA vaccine delivered by novel skin electroporationdevice protects mice against intranasal poxvirus challenge. Vaccine 2007;25(10): 1814–1823.

[179] C.Dobano, G.Widera, D.Rabussay, D.L.Doolan, Enhancement of antibody and cellular immune responses tomalaria DNA vaccines by *in vivo* electroporation. Vaccine 2007;25(36): 6635–6645.

[180] M.Wang, O.Orwar, S.G.Weber, Single cell transfection by electroporation using an electrolyte/plasmid-filled capillary, Anal Chem 2009;81: 4060-4067.

[181] J.Rathenberg, T.Nevian, V.Witzemann,High-efficiency transfection of individual neurons using modified electrophysiology techniques J.Neurosci Methods 2003;126: 91-98.

[182] Y.Saheki, S.T.Li, M.Matsushita, Y.M.Wu, W.H.Cai, F.Y.Wei, Y.F.Lu, A.Moriwaki, K.Tomizawi, H.Matsui, A new approach to inhibiting astrocytic IP3-induced intracellular calcium increase in an astrocyte-neuron co-culture system, Brain Research 2005;1055(1-2): 196-201.

[183] K.Hass, K.Jensen, W.C.Sin, L.Foa, H.T.Cline, Targeted electroporation in Xenopus tadpoles in vivo from single cell to the entire brain, Differentiation 2002;70: 148–154.

[184] S.M.Schanuel, K.A.Bell, S.C.Henderson, A.R.McQuiston, Heterologous expression of the invertebrate FMRFamide-gated sodium channel as a mechanism to selectively activate mammalian neurons, Neuroscience 2008;155(2): 374-386.

[185] K.Hass, W.C.Sin, A.Gavaherian, Z.Li, H.T.Cline, Single cell electroporation for gene transfer in vivo, Neuron 2001;29: 583-591.

[186] B.Judkewitz, M.Rizzi, K.Kitamura, M.Hausser, Targeted single cell electroporation of mammalian neurons in Vivo, Nature Protocals 2009;4: 862-869.

[187] M.Kanduser, D.Miklavcic, Electrotechnologies for extraction from food plants and Biomaterials, E.Vorobiew, N.Lebovka(eds), DOI: 10.1007/978-0-387-79374-0-1, Springer Science + Business Media, LLC 2008.

[188] J.Olofsson, K.Nolkrantz, F.Ryttsen, B.A.Lambies, S.G.Weber, O.Orwar, Single cell electroporation, Current Opinion in Biotechnology 2003;14: 29-34.

[189] D.S.Hewapathirane, K.Haas, Single cell electroporation in vivo within the intact developing brain, J.Vis.Exp.17,e705(2008) DOI : 10.3791/705 (2008).

[190] D.Karra, R.Dahm, The Journal of Euroscience 2010;30(18): 6171-6177.

Chapter 3

Flexible Circuit Technologies for Biomedical Applications

Damien C. Rodger, Wen Li, James D. Weiland, Mark S. Humayun and Yu-Chong Tai

Additional information is available at the end of the chapter

1. Introduction

The human body is an incredibly complex organism that exhibits an impressive resilience to external influences, such as infecting bacteria and inhaled particles. This, however, proves to be a difficult problem to overcome when designing and implementing implantable devices for medical therapy, as the body is supremely primed to resist any artificial intervention. One of the critical parameters to consider is the mechanical compatibility of the implanted device with the tissue of interest. Additionally, the footprint of the device should ideally be as minimal as possible to decrease tissue damage and to minimize encapsulation responses. Until recently, however, it has proven difficult to integrate naturally inflexible solid state circuits with flexible components due to connection complexities and pitch limitations. In addition, material and fabrication limitations have prevented the implementation of thin-film cables, radiofrequency coils, and neural stimulation electrode arrays in implantable devices.

Here, after a brief review of the literature, we discuss novel flexible technologies being developed for biomedical applications, and especially for high density functional neural stimulation and recording. The distinctive parameters of a unique material, namely the semicrystalline thermoplastic parylene C, that make it particularly well-suited as a biocompatible substrate for thin-film biomedical circuits are discussed, explaining the reasoning for its use in a novel paradigm as a substrate and not just as a coating. In addition, we discuss a new packaging scheme that has been developed to enable high lead count interconnects using microfabrication equipment for alignment and patterning, and demonstrate a complete parylene-based stimulation microsystem combining radiofrequency coils with solid state circuits and electrodes in a parylene substrate. This integration of solid state circuits with flexible, biocompatible components gives rejuvenated hope for a new generation of neural prosthetics aimed to provide eyesight to the blind and limb movement to para- and quadriplegic patients through spinal cord stimulation. These technologies can also enable integration of solid state devices with novel bioMEMS sensors in such a manner that devices previously only conceived of are now possible.

2. Outer retinal disease

Blindness due to such outer retinal diseases as retinitis pigmentosa (RP) and age-related macular degeneration (AMD) affect hundreds of thousands of people worldwide. In fact, it has been estimated by The Eye Diseases Prevalence Research Group and the National Eye Institute that AMD alone will affect three million people in the United States by the year 2020 [1]. In addition, the prevalence of RP has been estimated to be approximately 1 in 4000 [2]. Although the pathogeneses of these retinal photoreceptor diseases are, to date, not entirely understood, it is known that they are largely restricted to the outer retina, and that downstream circuitry, although it undergoes significant remodeling, is relatively spared [3-5]. There exist several possible approaches to thwarting the devastating effects of these diseases. Of these, the surgical, pharmacological, stem cell, and dietary approaches are promising. Laser ablation of leaky blood vessels and diets rich in antioxidants, for example, have been shown to slow the progress of AMD, but not to have any effect on its incidence. Effective pharmacologic agents, likewise, have long been elusive. Recent evidence has shown that stem cell therapy could also be a possibility for the treatment of such diseases, by possibly replacing the lost photoreceptors with stem cells capable of maturing into photoreceptors that then make connections with the rest of the retina [6]. However, such treatments, in reality, are still very far away from being used in clinical practice, and have a number of ethical and political barriers to their implementation.

In need of another possible treatment for these profound retinal diseases, in 1994 Humayun *et al.* reported the results of a pioneering study to electrically stimulate vertebrate retinas [7], a bioengineering approach that would possibly bypass the diseased photoreceptors in the visual pathway. In this study, bullfrog eyecups, as well as rabbit eyes (the rabbits were injected intravenously with sodium iodate, a chemical toxic to the retinal pigment epithelium and with other effects on photoreceptors), were stimulated with platinum electrodes. In their report, the following conclusion was made: "Surface electrical stimulation of the inner retina in normal eyes and in eyes with outer retinal degeneration can elicit a localized retinal response." This has since had a profound impact on the field of ophthalmology, by dramatically throwing an engineering solution into the mix of possible treatments for patients with devastating blindness due to outer retinal disease.

As a follow-up to this study, it was shown in 1996 by Humayun *et al.* that this seemingly simple approach of passing a carefully controlled electrical current through an electrode placed directly on the retinal surface to activate the still functional electrically excitable cells of the retina can also elicit visual percepts in humans with otherwise bare or no light-perception vision [8]. Perhaps even more important than the perception itself was the nature of the perception. Subjects did not report seeing streaks or multiple percepts of light simultaneously when stimulating from one electrode, as might be expected if the axons of the ganglion cells, as they made their way to the optic nerve in the nerve fiber layer, were being stimulated. Instead, they reported seeing discrete, highly localized percepts that seemed to correlate very well with the location of stimulation. This suggested that it was not the axons that were the most electrically excitable, but instead the cell bodies of the several retinal ganglion cells (RGCs) or the bipolar, or even possibly amacrine or horizontal cells underlying the electrode. Indeed, it has since been conjectured that it could even be the characteristic right-angle bend in the axon of the RGCs that has maximal electrical excitability [9]. No matter the case, this was the confirmation of the feasibility of an engineering approach to treat human blindness with a "retinal prosthesis."

There are now a large variety of approaches to artificial vision, each with their own advantages and disadvantages. There is this epiretinal approach (Figure 1), in which an electrode array is placed directly on the retina from its anterior aspect. There is the subretinal approach, in which an electrode or photodiode array is placed within the layers of the retina [10], as is also shown in Figure 1 (this can be further subdivided to the *ab interno* and *ab externo* approaches, in which an incision is made within the retina or within the sclera, choriocapillaris, or RPE to insert the device, respectively). There is also the optic nerve approach, in which electrodes are placed around the optic nerve in an attempt to coarsely stimulate the afferent ganglion cell axons [11]. Finally, there are also the cortical approaches, whether they be to the visual cortex [12] or to the lateral geniculate nucleus (LGN) of the thalamus [13, 14], the first stop for information from the retina as it enters the cortex. A complete discussion of the advantages and disadvantages of each of the different approaches is beyond the scope of this text, but can be found in several good review articles on visual and retinal prostheses.

The harbinger of an epiretinal prosthesis was the successful demonstration of a prototype 16-electrode device, fabricated by Second Sight Medical Products, Inc. (Sylmar, CA, USA) in six patients [15]. While clearly not enabling such activities of daily living as newspaper reading and facial recognition, previously completely blind subjects can, for instance, differentiate between a plate, a cup, and a knife, in a high-contrast environment free from background distracters, an undeniably incredible feat of engineering and medicine. Furthermore, subjects have been shown to be able to discriminate direction of movement of parallel white bars on a black background, and can locate white squares within a quadrant of otherwise black space [16]. This implant demonstrated the technology as well as the remarkable ability for the human brain to compensate for low-resolution input. Although this had previously been demonstrated by cochlear prostheses for patients with severe hearing impediments [17], it was unclear whether this plasticity would translate well to visual prostheses. In fact, patients have demonstrated their ability to discriminate large letters simply by *instinctively* scanning the camera mounted on their head back and forth over the image displayed in front of them, a sort of innate edge-detection mechanism, dramatically increasing the capability of the 16-electrode device on their retina. Few scientists disagree, however, that increasing the number of electrodes on the implanted array will dramatically add to the capabilities of the prosthesis as a whole in enabling the patient to better carry out activities of daily living. The prototype device is hand assembled, a fact that limits the resolution possibilities of the device. Indeed, were it still necessary to interconnect integrated circuits by hand, computers would be far less useful to us as they are today. The need exists, then, to bring microelectronics and microfabrication technology to bear on the problem of retinal prosthetic devices. The race is on to build high-density multielectrode arrays in such a way and with such materials that the method is scalable to the needs of long term, high-density retinal stimulation. Recently, a sixty electrode version of the epiretinal prosthesis has been tested and has gained the CE Mark in Europe for sale for the treatment of retinitis pigmentosa. The FDA is now reviewing the results of these same studies for potential approval in the United States as a therapy for outer retinal blindness.

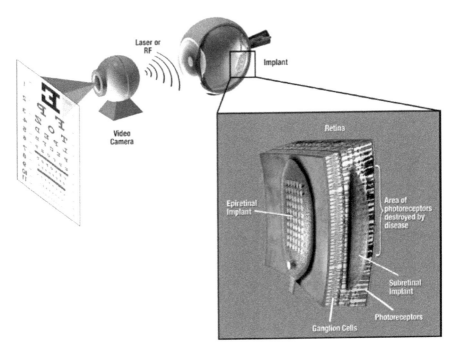

Figure 1. System overview and relative locations of epiretinal and subretinal implants [18].

3. Spinal cord injury

Spinal cord injury (SCI) can occur through a variety of mechanisms. The primary modalities fall into two major categories: trauma (e.g., automobile/motorcycle accident, sports injury such as from diving or horse riding, violence, fall), or disease (e.g., spina bifida or tumor). Spinal cord injuries can affect motor function, sensation, and autonomic functions (e.g., bladder control, breathing). Approximately 50% of SCIs in the United States are classified as complete [19]. This means that the spinal cord has lost the ability to transmit information across a segment within it, cutting off all functional communication from the brain to the nerves below the injury site and resulting in no sensation or voluntary control of motor function below the injury site due to lost input from the brain. This usually results in para- or quadriplegia, depending on the site of injury. Although in most cases the cord is not completely transected or even cut, it is significantly damaged by interruption of blood flow supplying one of its segments or through spinal contusion. Even though the vertebral column protects the cord, when trauma is sufficient to compromise this protective cage, the broken vertebrae can impinge on the cord and crush or destroy the axons within it very quickly, with continued loss of axons over time [20]. Some estimates pin the prevalence of spinal cord injury in the United States at approximately 250,000 [21], with an incidence of approximately 10,000 to 12,000 per year [22], while others state the prevalence is significantly higher, at around 450,000 [19]. Approximately 55% of spinal cord injuries occur in young victims between 16 and 30 years of age, making it a disease to bear usually for the rest of their lifetime, and more than 80% of victims are males [22]. In some cases of incomplete injury, function can be recovered over time [23]. However, in most cases, some level of impairment is permanent.

The complexity of the spinal cord is undeniable. In fact, despite the popular misconception that the spinal cord serves only as a communication conduit between the brain to the muscles and organs and from the skin back to the brain, it is much more accurate to view the spinal cord as an extension of the brain. While the spinal cord below a complete injury does indeed lose input from the motor cortex, and its ability to send sensations of touch to the brain is completely compromised, the spinal cord is not rendered useless. Even an intact cord does a lot of the primary processing and reflex control without any input to or from the brain.

In patients with spinal cord injury, there are several approaches to rehabilitation. The foot-drop stimulator, wherein stimulation of the peroneal nerve affects localized contraction of ankle dorsiflexors to counteract the problem of foot drag, has been widely studied [24] with mixed results [25-27]. Other functional movements requiring much more coordinated musculature responses are far more difficult with implantable peripheral nerve or muscle stimulators because of the need to control timing as well as pulse amplitudes of likely a large number of electrodes in rapid succession. Skin surface electrodes such as those in the ParaStep system [28, 29] suffer from these problems as well as the problem that many muscle groups are difficult to target from this more remote location. In addition, because in complete SCIs the voluntary input from the motor cortex to initiate such movements is lost, an accessory mechanism for determining the desired motion is warranted. This may require recording electrode arrays in the motor cortex as another component of this system, as well as possible electrical stimulatory feedback (in addition to the visual feedback already present). Any such system, then, is likely to be quite complex and difficult to implement in practice.

We propose a system, which eventually will be completely implantable, that is capable of stimulating the dorsum of the spinal cord in such a manner that modulation of the sensory input to the cord, interneuronal activity within, and even motor output from the cord, is possible. Tonic, subthreshold stimulation, applied at precise times and precise locations along the spinal cord, would likely help sustain or stop locomotor activity with the kind of coordination and rhythmicity already discussed. Perhaps in combination with both locomotor and standing training as well as appropriate pharmacological administration (e.g., quipazine), it is possible that such an array would give those with both complete and incomplete spinal injury the ability to stand and walk once again. In order to provide appropriate proprioceptive input in the case of complete SCI, it would likely be possible to provide an accessory device or muscle stimulator to initiate this type of activity. However, this approach leverages the innate activity and processing power present in the spinal cord to its greatest extent and likely obviates the need for a stimulation control system of great complexity in order to bring about coordinated muscle activity, as is necessary for a peripheral muscle or motor neuron control system. Such electrical stimulation may also, as has been hypothesized in the case of subthreshold retinal stimulation systems as well, have the capability of promoting axon regrowth [30, 31] and facilitating plastic changes in the cord. Such an array would likely need to be conformable to the cord, and would need to be implanted in relatively close apposition to it, either epidurally (from which location the electrical field would need to penetrate several meningeal layers) or subdurally, such that it is closer to the surface of the cord. In the ideal case, the array would be flexible enough to move with the cord during motion and bending such that functional targets are the same in any position. In addition, it should be the case that such an array can also record from the cord so as to recognize returning action potentials from the dorsal root and modulate this input accordingly. This approach, then, requires a high-density array with many electrode sites from which to choose during training and daily activity, as well as the ability to both record from and stimulate the cord, something which current arrays, such as those for pain management, simply can not do because they are too bulky, inflexible, and of too low a density (up to 16 electrodes).

4. Introduction to parylene

Parylene is the trade name for a family of semicrystalline thermoplastic polymers known as the poly(para-xylylenes) (PPX). These were discovered in 1947 by Michael Szwarc in Manchester, England [32]. It was originally deposited in an investigation of aliphatic carbon-hydrogen bonds where the carbon was attached directly to a benzene ring, by heating toluenes and ortho-, meta- and para-xylenes to very high temperatures and looking for degradation products [33]. The very first parylene film deposited, then, was parylene N (with no substitutions on the benzene ring), and was temporarily known as a Szwarcite snakeskin. This deposition process, however, had the disadvantage that gaseous by-products were an inherent part of the method. William Gorham, an employee at Union Carbide, soon thereafter devised an alternate method of parylene deposition that involved the pyrolysis of a dimerized form of the material, di-para-xylylene, or [2.2] paracyclophane. It was largely unknown at the time how to make this dimer in large amounts as it had only been isolated as an impurity in Szwarc's method. However, in 1951 Donald Cram reported a method for making this material in bulk. On February 17, 1965, Union Carbide announced the availability of parylene films and the new vacuum deposition method, known as the Gorham process. There were over 20 types of parylene actually developed, but only three were considered commercially viable: these were parylene N (no chlorines on the benzene ring), parylene C (one chlorine on the benzene ring), and parylene D (with two chlorines on the benzene ring) (Figure 2). However, a new fluorinated version of parylene, parylene HT, has recently become commercially available, and can be deposited in a new parylene deposition system.

In the Gorham vapor-deposition process [34] (depicted in Figure 3), which takes place at vacuum (~25-35 mT, to increase mean free path to the substrate), a charge of parylene dimer is placed in a vaporizer furnace. The dimer evaporates at approximately 130 to 150 °C, and then passes through a very high temperature pyrolysis (~650 to 750 °C) furnace, where the molecule is split into monomers. The monomers enter a chamber that is held at room temperature, and polymerize on all exposed surfaces in the chamber conformally and without pinholes. Residual monomer is collected on a cold trap. Different variants of parylene require varying process conditions, but the method remains essentially the same.

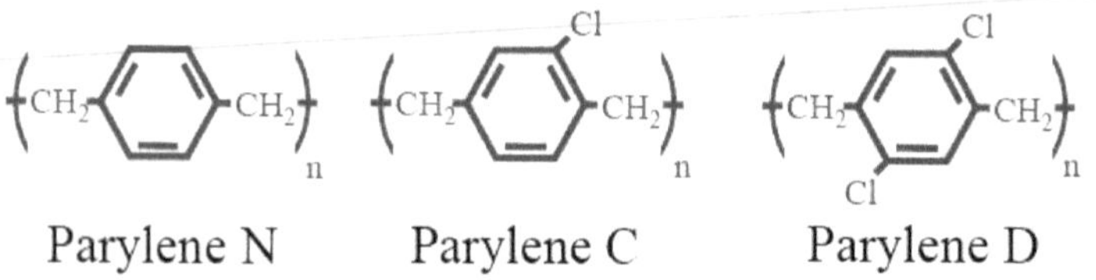

Parylene N   Parylene C   Parylene D

Figure 2. The three originally commercialized parylene variants.

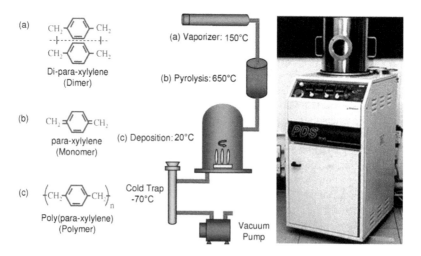

Figure 3. center) Gorham process for parylene deposition. (left) Chemical structures at corresponding points. (right) PDS 2010 Labcoater System.

The existing variants of parylene have varying properties as well as uses. Parylene N is primarily used as a dielectric and when lubricity and crevice penetration is important. Parylene C is an ISO 10993, United States Pharmacopeia (USP) Class VI material (the highest biocompatibility rating for plastics in the United States) and has excellent water barrier properties. In addition, it has a very large elongation to break. Parylene D has now been largely replaced by parylene HT [35], but is used when mechanical strength is of primary concern. Parylene HT has extremely low coefficients of static and kinetic friction, excellent thermal stability and good water barrier properties, high ultraviolet stability, and is also ISO 10993 biocompatible [35-37]. Finally, very recently, Kishimoto Sangyo Co., Ltd. in Japan has devised additional parylenes in which amino groups have been added to the benzene rings. The amino group may add even more improved biostability, but could also generate bioactivity that may or may not be beneficial to device functionality. We have preliminarily tested these amino parylenes have found them to be compatible with standard parylene processing technology, such as oxygen plasma reactive-ion etching (RIE), in our cleanroom, but they will not be explored further in this text.

The advantages of using parylene, and, more specifically, parylene C, as the structural material for neuroprostheses, when compared with technologies based on the use of other materials such as PDMS, polyimide [38] and silicon [39], include parylene's pinhole-free conformality due to its unique room-temperature chemical vapor deposition process, its low water permeability, its chronic implantability and its high flexibility and mechanical strength (Young's modulus ~4 GPa). The Young's moduli of two other commonly used materials for neuroprostheses, PDMS and polyimide 2611, are graphed alongside that of parylene C in Figure 4. PDMS arrays have been handled by surgeons in our vivarium, and these are often as reported as too floppy and difficult to handle due in part to the low Young's modulus, hence requiring very large thicknesses to handle appropriately. In addition, polyimide 2611 (often chosen because its water permeability is lower than that of other polyimides) has a Young's modulus larger than that of parylene C. It has been suggested that polyimide arrays often are too rigid and can damage the retina. In addition, they tear quite easily. Parylene C handles very well under surgical manipulation, and, as we will show, behaves very well when implanted. In addition, the thickness of parylene films is determined by the mass of dimer placed in the vaporizer. As such, thicknesses can be very thin or very thick, and thicknesses are very repeatable and well controlled, more so than spin-on coatings, especially when deposited over step junctions. Parylene thickness can be controlled so as to match the rigidity to the application of interest, more easily than most other materials. Since parylene is deposited at room temperature (we have verified this using Temp-Plate irreversible temperature recorders traceable to NIST (Wahl Instruments, Inc., Asheville, NC, USA)), the coating process is post-IC compatible. Parylene C is also optically transparent, enabling the anatomy to be seen through the cable and the array during ophthalmic surgery, post-implantation examination, and follow-up. While many groups use parylene C as a coating on their arrays for many of these reasons, we have chosen to use it as the main substrate for our devices [40, 41], a paradigm that leverages these advantages to the greatest extent.

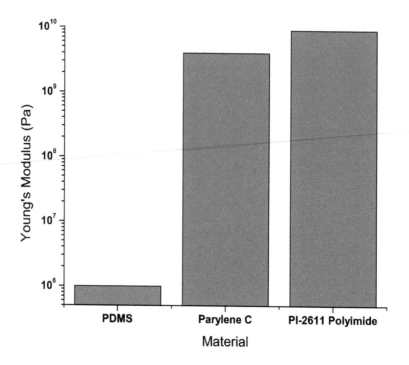

Figure 4. Comparison of Young's moduli of various materials. Ordinate is logarithmic.

Although parylene is known to be biocompatible in many sites of implantation, in order to initially assess the intraocular biocompatibility of the material in the unique immune environment of the eye, an approximately 2 cm × 0.5 cm piece of unmodified 20 μm thick parylene C was implanted in the vitreous cavity of the right eyes of two rabbits for six months. The retinas of the right eyes of both rabbits were compared post-mortem with those of their left eyes that served as controls. Histological evaluation revealed no discernable difference between right and left eyes, indicating that there was no detectable adverse retinitis, choroiditis, endophthalmitis, or scleritis seen as a result of parylene implantation in the vitreous cavity [40]. These results supported the tenet that parylene C is a biocompatible bulk material for an intraocular retinal prosthesis and other ocular implants, and paved the way for the design and fabrication of a flexible electrode arrays and a packaging system using parylene C as the primary substrate. Given these biocompatibility results, we have also been investigating parylene in several other ocular implants, with excellent results to date [42-44]. Similar experiments were performed with parylene C implanted on the spinal cord of mice. The arrays were well tolerated, with no obvious immune reaction or gliosis.

5. Single-metal-layer flexible MEAs

5.1. Fabrication

Single-metal-layer parylene C-based electrode arrays are fabricated as shown in Figure 5. A photoresist sacrificial layer is optionally spun on a standard silicon wafer. Approximately 8 μm of parylene C is then vapor deposited in a PDS2010 system (Specialty Coating Systems, Indianapolis, IN, USA) on the entire wafer. An LOR3B photoresist layer (Microchem Corp., Newton, MA, USA) and an AZ1518 layer (AZ Electronic Materials, Branchburg, NJ, USA) are spun on top of the parylene, exposed in a 10X reduction GCA Mann 4800 DSW wafer stepper (General Signal Corporation, Stamford, CT, USA) or a Kasper 2001 contact aligner (Kasper Instruments, Inc., Sunnyvale, CA, USA) depending on the required resolution of the electrode array, and developed to achieve a liftoff pattern comprising contacts, conductive traces, and electrodes. After hard bake, approximately 2000 Å to 5000 Å of platinum, with or without a 200 Å titanium layer, is then e-beam evaporated (SE600 RAP, CHA Industries, Fremont, CA, USA) on the wafer. The subsequent photoresist strip generates the desired single-layer metallization pattern. An approximately 7 μm thick coating of parylene C is then deposited, followed by a spin coating of photoresist. This photoresist etch mask is exposed over the areas of the electrodes and contact pads and to pattern the overall array geometry, and the entire wafer is then subjected to an RIE in oxygen plasma, removing the parylene insulation over the electrodes and the parylene surrounding the array. The photoresist mask is then removed with solvent. Finally, if a sacrificial photoresist layer was used, the array is released from the substrate in an acetone bath. If no sacrificial layer was used, it is peeled from the silicon in a water bath. Ultimately, for most cases, the sacrificial photoresist layer is unnecessary, and can often complicate array fabrication due to cracking while under process. The arrays can be easily released from a natively oxidized silicon surface by placing them in a deionized water bath and peeling them from their edge. The water will then release the rest of the structure due to the hydrophobicity of the underlying parylene surface.

A single-layer square-grid electrode array, consisting of 256 Ti/Pt thin-film electrodes 125 μm in diameter in a 16 × 16 grid with connecting lines of 12 μm pitch fabricated in the manner of Figure 5 is shown in Figure 6. An SEM highlighting the typical electrode morphology in such structures is shown in Figure 7. As can be seen, the parylene covering the electrode has been completely removed, whereas the incoming trace remains conformally coated with the material.

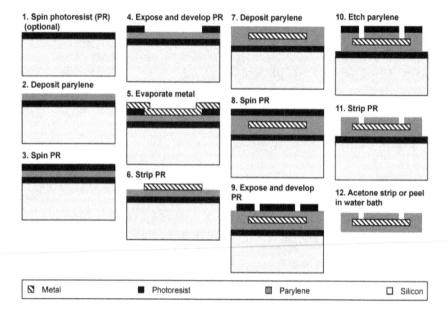

Figure 5. Fabrication process for parylene-based single-metal-layer flexible MEAs.

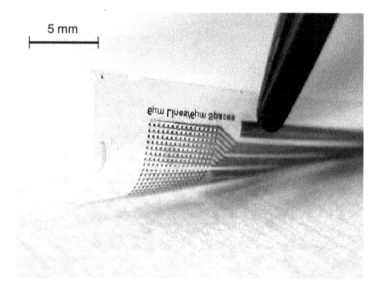

Figure 6. Photograph of Ti/Pt electrode array of 256 electrodes and lines of 12 μm pitch.

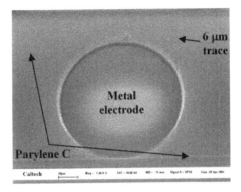

Figure 7. SEM of electrode morphology showing parylene C insulation surrounding exposed metal electrode.

5.2. *In vitro* retinal recording and stimulation

Parylene C-based arrays of thin-film platinum electrodes, comprising four 200 μm diameter stimulating electrodes and 56 recording electrodes of 10 μm diameter were fabricated according to the single-metal-layer process on a glass substrate. These were placed in a bicarbonate perfusate under a microscope and connected to a stimulus generator and preamplification board (Multi Channel Systems MCS GmbH, Reutlingen, Germany) [45]. As shown in Figure 8, a retina isolated from larval tiger salamander (*Ambystoma tigrinum*) was placed RGC side down on the array (to simulate epiretinal stimulation), and a remote platinum ground electrode was introduced to the bath.

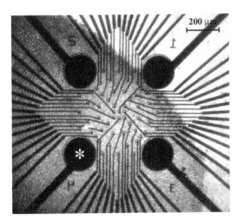

Figure 8. Isolated larval tiger salamander retina (darker region at left) overlying parylene-based platinum electrode array. Arrow indicates 10 μm diameter electrode used for recording trace in Figure 9. Asterisk identifies 200 μm diameter stimulating electrode used to generate action potentials seen in Figure 9.

With the lights off, a 20 μA, 400 μs/phase, cathodic-first biphasic electrical pulse was applied between the stimulating electrode indicated with an asterisk in Figure 8 and the ground electrode. The voltage trace from the recording electrode is shown in Figure 9. This stimulation was consistently repeatable over a 50 pulse train with a 400 ms inter-pulse interval, and other stimulating electrodes were also capable of "epiretinally" stimulating other cells in the retinal slice. As is clear from these results, the parylene-based platinum electrode was able to stimulate the tissue and elicit a response similar to the response generated from a light pulse in this intact retina. Given these results and the knowledge garnered from clinical trials with prototype arrays fabricated of other materials, it is not unreasonable to presume that our arrays will most likely be able to stimulate retinal tissue in other species, including human.

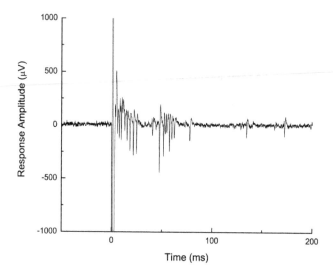

Figure 9. Typical recording of response of cells overlying recording electrode to a 20 μA, 400 μs/phase, cathodic-first biphasic electrical pulse from "epiretinal" stimulating parylene-based platinum electrode denoted with an asterisk in Figure 8.

6. Spinal cord prosthesis

6.1. System overview

The ideal spinal cord stimulation system, just like the retinal system, would have a power source, circuitry for driving the appropriate electrodes, as well as a cable and electrode array, this time implanted epidurally or subdurally on the spinal cord. We believe that a penetrating electrode array would be problematic for implantation and would likely lose efficacy and fail ultimately due to a gliosis over time, as has been shown in many other studies [27]. The power source could be an RF coil, or could, due to the much larger space available in the abdomen and back as compared with the eye, be a rechargeable battery capable of charging through the inductive link. The RF coil, in addition, would enable reprogramming of the implanted electronics for alternative stimulation protocols at the physician's discretion. The electrode array should be conformable to the spinal cord so that it can tonically stimulate at low currents and with high precision. While a completely implantable system is the ultimate goal, an interim goal is to stimulate the spinal cord chronically from an array connected to a head plug, while simultaneously being able to record electromyograms (EMGs). In order to achieve this, we have studied the efficacy of the multielectrode array portion of this system and have begun to develop a connector technology capable of connecting 36 electrodes in with a small enough form factor to be chronically mounted on a mouse skull.

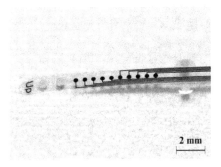

Figure 10. Parylene MEA for murine spinal cord stimulation and recording.

6.2. Fabrication

Spinal cord arrays, consisting of five or ten electrodes of 250 μm diameter were designed and fabricated (Figure 10). Interelectrode spacing was controlled so that each array of electrodes covered four to five segments of the murine lumbosacral spinal cord upon implantation. Suture holes were also designed into the body of the array to ensure placement and attachment of the array on the cord, as well as to facilitate implantation (suture can be attached to the end of the array and can be threaded along the cord first to help direct the array along it).

The single-metal layer fabrication process was performed using a contact aligner process for fast throughput. The fabricated arrays were annealed to increase the adhesion of parylene to parylene. At the same time, they were clamped between two pieces of Teflon or glass slides coated with aluminum foil to ensure they would be flat during implantation. The arrays were connected via Clincher connectors (FCI, Versailles Cedex, France) to the stimulation and recording electronics.

6.3. Implantation and testing

Just prior to implantation, the arrays were rinsed in isopropyl alcohol. Under isoflurane anesthesia, the spinal cord electrode arrays were implanted epidurally on spinal cord segments L2-S1 in nontransected mice. The electrodes were oriented linearly along the rostrocaudal extent of the cord. Recording capability was assessed by using the electrode array to record spinal cord potentials evoked by tibial nerve stimulation. Following stimulation of the tibial nerve, somatosensory evoked potentials were recorded from the cord dorsum at three lumbosacral levels (P1-P3, rostral to caudal). The recorded waveform consisted of three response peaks, two of which are clearly depicted in Figure 11 (N1 and N3). These findings closely mirror results reported previously in a study using conventional spinal cord recording electrodes [46] demonstrating that the recording capability of the array electrodes matches that of conventional electrodes. By measuring the difference in the response latencies obtained at each electrode position (corresponding to different levels of the spinal cord), and by utilizing the known, fixed interelectrode spacing, accurate measurements of the conduction velocities were obtained. The properties of these responses can potentially be used to diagnose the progressive recovery of the spinal cord as a result of treatments provided after a spinal cord injury.

To test the capability of the electrode array to act as a multichannel stimulating device for generating hindlimb movements, constant-current monophasic stimulus pulses (amplitude: 50-850 µA, frequency: 0.3-10 Hz, pulse duration: 0.5 ms) were applied to the spinal cord between each of the array electrodes and a ground electrode located near the shoulder, while muscle activity was monitored using electromyogram (EMG) recordings of the tibialis anterior and medial gastrocnemius muscles. Stimulation generated a typical three-component EMG action potential consisting of an early (direct motor), a middle (monosynaptic), and a late (polysynaptic) response, classified by post-stimulus latency (Figure 12). These data clearly indicate that the parylene arrays were able to stimulate the spinal cord in such a way that the musculature was activated.

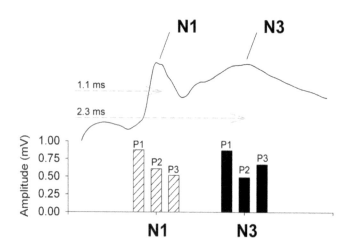

Figure 11. Peak amplitudes of somatosensory evoked potentials (N1 and N3) recorded from three levels of the rostrocaudal spinal cord (P1-P3). Example waveform at top shows approximate response times.

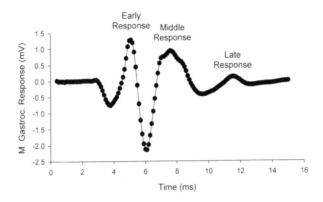

Figure 12. Typical medial gastrocnemius (ankle plantarflexor) EMG recording showing early, middle, and late responses after stimulation of spinal cord with parylene MEA.

Because of the known spacing of the electrodes on the array (as compared with traditional fine-wire electrodes which do not have known interelectrode spacing), we were able, in addition, to determine whether electrode position had a significant impact on muscle recruitment. The appearance and magnitude of each of the EMG responses was indeed correlated with the choice of electrode position (Figure 13). This serves as evidence that position of stimulation is very important. With a one-dimensional array, it is difficult to assess whether a bilateral stimulation paradigm would also result in lateralization of response, but we strongly suspect that this would be the case.

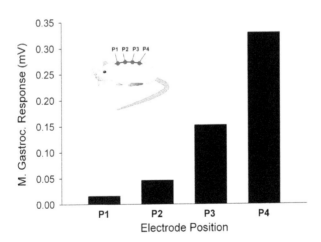

Figure 13. Medial gastrocnemius EMG showing varying levels of activation due to stimulation at different rostrocaudally located electrode sites.

7. Dual-metal-layer flexible MEAs

7.1. Motivation

An inherent problem with single-metal-layer arrays is that electrodes and traces are necessarily fabricated alongside each other. For high-lead-count devices, this limits the size of electrodes and tends to crowd electrodes and traces into artificial groups. To allay these problems, a dual-metal-layer approach was devised that enables traces to pass underneath overlying electrodes.

7.2. Fabrication

Dual-metal-layer electrode arrays are fabricated as shown in Figure 14. Approximately 8 μm of parylene C is first deposited on a silicon wafer with the optional photoresist sacrificial layer, forming the underside of the electrode array. A platinum or titanium-platinum metal liftoff process is used to define traces with 16 μm pitch and 2000 Å to 3000 Å thickness. A second parylene deposition (~1 μm) forms the insulation between the two metal layers. At this point, 6 μm by 6 μm vias are patterned in the insulation layer over the ends of the traces using an O_2 plasma RIE. A second step-coverage optimized liftoff process is used to define a second metal layer comprising electrodes and traces, while at the same time achieving electrical continuity between the underlying traces and the overlying electrodes. A final parylene coating approximately 7 μm thick forms the top insulation. The electrodes are exposed and the overall geometry of the implant is defined in a final set of O_2 reactive-ion etches using a thick photoresist etch mask. Finally, the arrays are peeled from the wafer in a water bath or released through removal of the sacrificial photoresist in acetone. The process depends on optimal step coverage of the parylene sidewall during evaporation, which is aided, in part, by the slightly isotropic nature of the O_2 plasma etch of parylene [47] as well as by the special design of the rotating wafer domes inside the e-beam evaporator, for which the angle of attack of the metal evaporant is adjusted for best coverage. This requirement conflicts marginally with those for successful metal liftoff, however, in this case, the liftoff technique is robust even under these step-coverage optimized conditions due to the choice of an LOR/positive photo-resist compound layer.

7.3. Implantation and testing

Chronically implantable retinal electrode arrays comprising 1024 75 μm diameter electrodes arranged in a complex biomimetic pattern that closely mimics the density of RGC in the human retina [48] were designed (the electrode density varied radially in a ratio matched to that of the RGCs), as shown at the left in Figure 15. These arrays (shown at right in Figure 15) were fabricated according to the dual-layer process, with 60 of the electrodes connected via two traces each to facilitate electrical conductivity verification. The strength of metal adhesion was verified using a Scotch tape test, which demonstrated that direct platinum evaporation is feasible without the need for a titanium adhesion layer. Electrical testing demonstrated a typical line impedance of a contact-electrode-contact circuit to be approximately 5 kΩ, which included two 8 μm wide traces of 20 mm length, as well as two via step junctions connecting

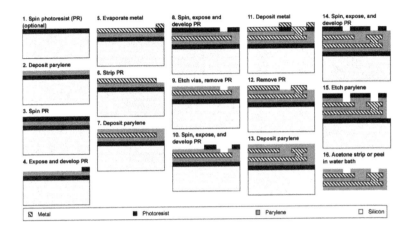

Figure 14. Fabrication process for parylene-based dual-metal-layer flexible MEAs.

underlying traces to the overlying electrode. Two types of via and electrode configurations were tested. Some electrodes (electrode SEM given in Figure 16 (left)) had vias connecting to the underlying trace near the center of the electrode, hence enabling charge spreading from the center of the electrode. One possible drawback to this configuration is that the contact from trace to electrode over the sidewall is a potentially vulnerable point of the circuit during processing (e.g., subsequent RIE processes) and during pulsing in electrolyte because the metal may be thinner there. The other electrode configuration had vias located adjacent to the electrode (electrode SEM given in Figure 16 (right)), with the possible advantage that it would be protected during RIE and subsequent pulsing by the overlying conformal parylene layer. An SEM showing the morphology of a single central via is given in Figure 17. This clearly depicts the sidewall coverage responsible for the electrical contact between the first and the second metal layers. In both configurations, each via had an impedance of less than 12.5 Ω. The best final configuration has not yet been determined.

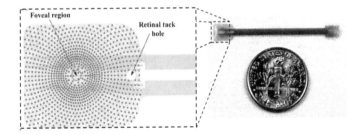

Figure 15. (left) Design of biomimetic dual-metal-layer retinal electrode array showing biomimetic arrangement of electrodes. (right) Fabricated biomimetic chronically implantable arrays with 60 of 1024 75 µm diameter electrodes connected through dual-layer process with U.S. dime for size comparison.

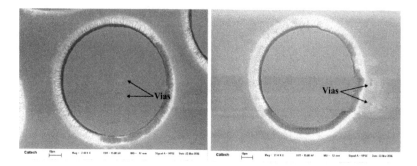

Figure 16. Two possible dual-layer electrode configurations. Electrode with central vias (left), and electrode with abutting vias (right).

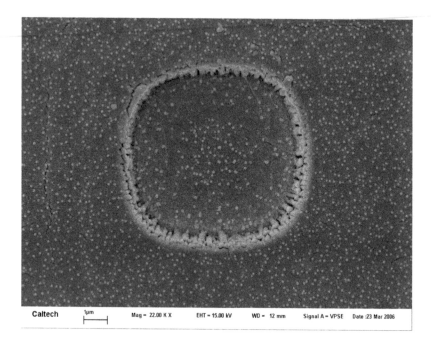

Figure 17. Magnified view of trace to electrode via showing sidewall coverage.

The arrays were successfully molded to the approximate curvature of the canine retina (Figure 18 (top)) using heat-annealing and a custom mold in a vacuum chamber, and sterilized using ethylene oxide gas. Two biomimetic arrays were implanted in the right eye of two canines through a 5 mm pars plana incision after vitrectomy, and were affixed to the retina (Figure

18 (bottom)) using a retinal tack modified by the addition of a PDMS washer (to account for the thin nature of the parylene arrays).

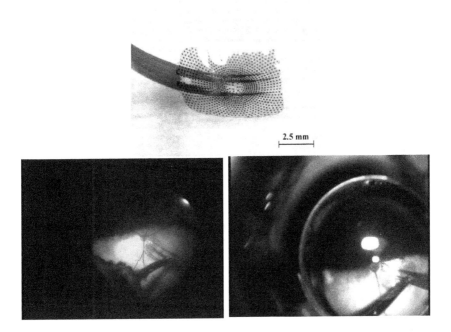

Figure 18. Heat-molded and annealed retinal electrode array with retained spherical curvature (top), and intraoperative photographs of tacking in each canine (bottom).

Follow-up in both animals was conducted for six months using fundus photography, fluorescein angiography (FA), in which blood is fluorescently stained to assess vessel perfusion in the retina, and optical coherence tomography (OCT), an interferometric technique that enables cross-sectional imaging of the retina. Fundus photography and FAs of both animals, examples of which are shown in Figure 19, consistently demonstrated that vessel filling underneath the array was normal. Obstruction and vessel leakage would have been visualized if the array were placing excessive pressure on the retina. In addition, OCT demonstrated that the electrodes were consistently less than 50 μm away from the ganglion cell layer in both animals

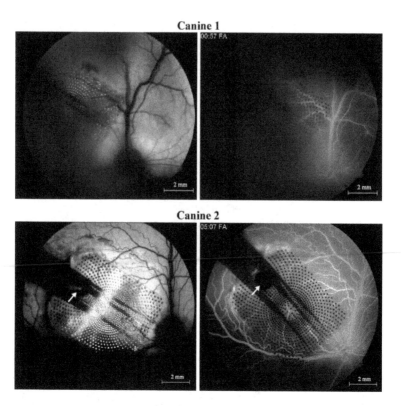

Figure 19. Fundus photographs (left) showing parylene MEAs tacked to the right retina of both animals and FAs (right) showing normal vessel perfusion under the arrays. Arrows point to retinal tacks.

(typical OCTs of both animals are shown in Figure 20), an outcome that theoretically would afford excellent electrical coupling between the electrodes and the electrically excitable cells of the retina. It is important to note that in the OCT of the second canine, the scan was taken along a segment furthest from the tack site, where one might expect the least proximity. Even at this location, this array remained in very close apposition throughout the six-month implantation. Post-enucleation histology has since confirmed the excellent biostability seen during follow-up.

As was briefly mentioned previously, the dual-layer process proffers considerable advantages over the more traditional single-layer approach. Design of single-layer electrode arrays is usually hindered by the need to route traces amongst the electrodes. This tends to cause crowding of traces and electrodes into groups, an organization that may not be optimal for stimulating the tissue of interest. In addition, this has a propensity to constrain the geometric area of the electrodes in the MEAs to smaller sizes, and thus reduces the number of electrodes possible in a given area. The dual-layer process obviates these problems by

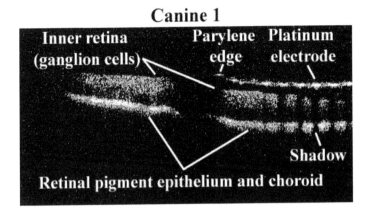

Figure 20. OCTs of both animals showing very close apposition (<50 µm) of the arrays to the RGC layer.

enabling traces to pass under overlying electrodes without making contact to them, having the effect of both relaxing the constraints on electrode size and number and enabling more complex electrode organization (such as the biomimetic one presented in this work). Although the arrays fabricated here had just 60 electrodes of connectivity with 120 traces total, this was without making full use of both layers for wire routing and connection of electrodes. In order to not make traces unnecessarily narrow and of too high impedance, we believe an extension of this process to three or more metal layers will be necessary to achieve 1024 electrodes of total connectivity. Indeed, this fabrication process is easily extendable to create such structures through addition of extra layers of parylene and metal. Given the encouraging biostability results presented here and the ability of these arrays to stimulate retinal tissue, future studies will include chronic stimulation from implanted parylene-based arrays in an animal model.

8. Electroplating

As a possible mechanism for extending the longevity of chronically pulsed electrodes, we have investigated electroplated films of high surface-area platinum. Specially designed thin-film platinum electrode arrays, consisting of sixteen 75 μm and 150 μm diameter electrodes of 3000 μm center-to-center spacing, were fabricated according to the single-layer process. Initial experiments were performed on these arrays to determine material morphologies after plating at different potentials in an aqueous ammonium hexachloroplatinate solution according to Whalen *et al.* [49]. Subsequently, arrays were immersed in the solution in a specialized jig and six were plated at a plating potential of -0.6 V (vs. an Ag/AgCl reference electrode) for 1.5 hours. The others remained unplated. Electrochemical tests were performed to evaluate the efficacy of this plating step in extending electrode longevity under chronic pulsing.

The electrode morphologies of a typical array of 16 electrodes plated at different potentials are shown in Figure 21. Note that the 4 corner electrodes (1, 4, 13, and 16) were not plated. Magnified views of some of the possible morphologies attainable via this mechanism are shown in Figure 22. These micrographs show morphologies that likely correspond to a drastically increased surface area. In order to confirm this cyclic voltammograms (CVs) in O_2-free H_2SO_4 were taken of the electrodes before and after platinization. According to [50], "real" electrode surface area can be adequately estimated by integration of the area over the hydrogen adsorption peak or likewise the area under the hydrogen desorption peak. The two CVs in Figure 23 show that there was a more than 40-fold increase in surface area after platinization when compared with the pre-plated surface area (note the change in scale of the ordinate from nA to μA). Under pulsing, the voltage responses of both plated and unplated electrodes remained stable for approximately 29 days, at which point the unplated electrodes showed signs of failure. Voltage responses for one such electrode on day 26, 29, and 31 are overlaid in Figure 24 (left), which documents the progression of failure. The plated electrodes, on the other hand, remained intact for much longer, most surviving more than 50 days, or 430 million pulses, at which point the testing goal was met and the test was stopped. Overlaid voltage responses for one such electrode, showing the voltage responses at day 26, day 31, and day 50, are shown in Figure 25 (right). The electrochemical impedances at 1 kHz of a typical plated and unplated electrode are shown in Figure 25. A dramatic jump in impedance was observed for the unplated electrode at the time of failure, while the plated electrode demonstrated only minor variability in its lower impedance throughout the 430-million-pulse trial (most variability happened on the days that CVs were taken, as expected). These preliminary data corroborate the evidence that plating of the electrodes is beneficial to longevity, and suggest that high surface-area platinization of electrodes can have a dramatic effect on extending electrode life while lowering electrochemical impedance to charge delivery. Future work will include replication of these tests and chronic pulsing at high temperatures for longer times to further accelerate and assess the possible modes of failure.

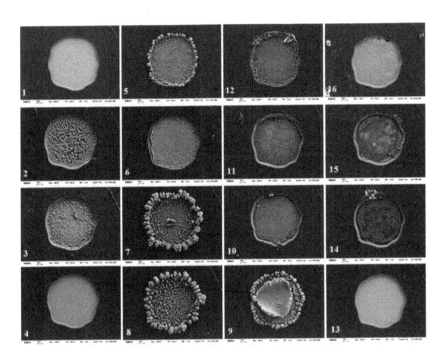

Figure 21. SEM of each of the electrodes on a typical 16-electrode array after platinization.

Figure 22. Magnified views of possible surface morphologies after platinization.

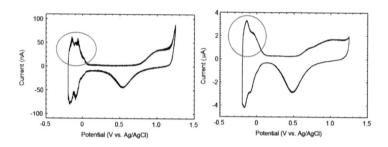

Figure 23. CVs showing more than 40-fold increase in electrode surface area from before plating (left) and after plating (right). The surface area is estimated by integrating the area under the peaks circled in red. Note change in scale of the ordinate. Scan rate: 100 mV/s. Electrolyte: O_2-free H_2SO_4 (N_2-bubbled).

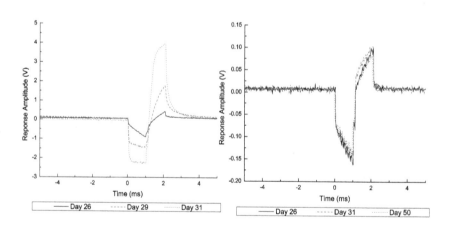

Figure 24. Voltage responses to a current pulse for (left) an unplated electrode, documenting the process of electrode failure, and (right) a plated electrode, showing steady responses throughout the 50 day test. Note response amplitudes for plated electrodes are far lower than those for the unplated electrode, as expected [27].

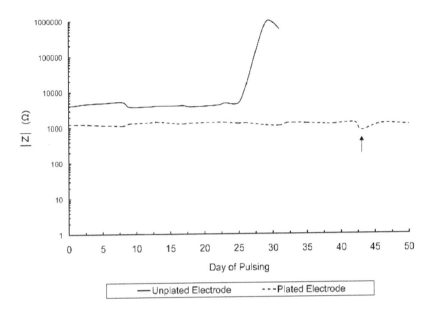

Figure 25. Magnitude of the electrochemical impedances at 1 kHz of an unplated and plated electrode over time. The unplated electrode showed a dramatic increase in impedance around day 30, at which time the test was stopped, whereas the plated electrode showed steady impedance through day 50. The arrow denotes a temporary dip in impedance due to CV scanning.

9. Chip-level integrated interconnect technology

9.1. Introduction

Despite our ability to fabricate such a large number of electrodes in such a small area, a significant impediment to future progress is the problem of how to package and interconnect these multielectrode arrays with foundry-fabricated ASICs, discrete components (e.g., chip capacitors, oscillators, diodes) and RF coils in a way that provides for high lead-count interconnects. A wafer-level process is cost prohibitive, as it is necessary to maximize the area of a wafer devoted to IC processing to keep costs low. Furthermore, current technologies for packaging would be far too tedious and low yield to apply to a 1000-electrode device. In order to achieve our goal of a 1000-electrode retinal prosthesis and a high-density spinal cord stimulation system, then, a new way of forming such a package so as to enable high-lead-count integration is necessary.

We have invented a way to place prefabricated chips, manufactured, for example, at a foundry, into the fabrication process of a parylene-based multielectrode array and/or RF coil, such that all interconnections to the chip are made using standard photolithography and standard microfabrication techniques in a fully scalable manner [40]. This packaging scheme is known as the chip-level integrated interconnect (CL-I^2) package. Figure 26 shows an overview of the fabrication process and how multiple chips could be joined together in this manner. A detailed discussion of the fabrication process, as adapted from [40], follows.

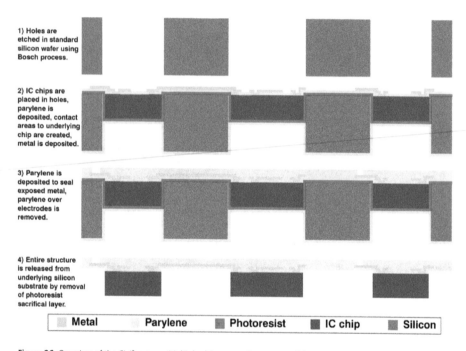

1) Holes are etched in standard silicon wafer using Bosch process.

2) IC chips are placed in holes, parylene is deposited, contact areas to underlying chip are created, metal is deposited.

3) Parylene is deposited to seal exposed metal, parylene over electrodes is removed.

4) Entire structure is released from underlying silicon substrate by removal of photoresist sacrifical layer.

| Metal | Parylene | Photoresist | IC chip | Silicon |

Figure 26. Overview of the CL-I^2 process. Multiple chip connections are possible.

9.2. Fabrication

Three MOSIS-fabricated ASICs, as well as seven chips fabricated to simulate them (with circuitry that facilitated testing), were used to demonstrate the CL-I^2 packaging technology. In order to fabricate the replicas of the MOSIS chips, these chips were imaged using a WYKO interferometer (Veeco Instruments Inc., Woodbury, NY, USA), and were found to have mean dimensions of 2.500 mm in length, 2.617 mm in width, and 254.2 μm in total thickness.

One hundred angstroms of chrome and 2000 Å of gold were e-beam evaporated on a 260 μm thick silicon wafer. Using a photoresist mask, the metal was wet etched to pattern pads of the same size and in the same locations as on the MOSIS-fabricated chips (approximately 70–100 × 100 μm^2 with a center-to-center pad spacing of approximately 200 μm), as well as a pattern of short circuits connecting these pads to nearby pads. After stripping the photoresist, a second photoresist layer was spun on the wafer and patterned as a mask for a Bosch through-wafer etch in a PlasmaTherm SLR-770B deep reactive ion etching (DRIE) system (Unaxis Corporation, St. Petersburg, FL, USA). This etch defined the length, width, and thickness of the simulated chips as 2.49 mm, 2.61 mm, and 260 µm, respectively. Finally, the photoresist mask was removed from the individual chips. In this manner, chips comprising simple electrical shorts and intrinsic resistors were fabricated as our primary CL-I² package test structures (Figure 27.)

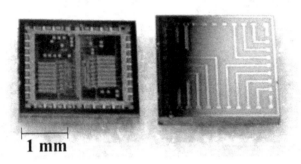

Figure 27. MOSIS ASIC (left) next to test chip (right).

The only properties of these prefabricated chips that had to be known before incorporation in the CL-I² process were their overall length, width, and thickness, and the dimensions and locations of the contact pads. Figure 28 gives a detailed CL-I² process flow. To begin, shallow alignment marks are etched into a standard 550 µm thick silicon wafer using a thin photoresist mask and an SF₆ plasma. 2.51 × 2.63 mm² holes are then patterned after alignment in a 10X reduction stepper in thick photoresist and an optional silicon dioxide mask. Through holes are then etched using the Bosch DRIE process. After photoresist and oxide removal, Nitto tape is placed on the frontside of the wafer. The chips are then self-aligned in the holes by inserting them from the backside (the Nitto tape enables frontside planarization whereas the lateral dimensions of the etched cavity determine lateral displacement), and they are sealed in place using several drops of sacrificial photoresist to cover the backside of the chip and to fill the gaps around it. A subsequent approximately 12 µm thick parylene C deposition in a PDS2010 mechanically anchors the chips in place from the backside. After removal of the frontside parylene by peeling off the Nitto tape, vertical displacements of the chips are measured using a stylus profilometer (Alphastep 200 and P-15, KLA-Tencor, San Jose, CA, USA).

The parylene-based flexible electrodes, or, in this implementation, contact pads for electrical testing, are then fabricated on this wafer as if it were a whole wafer with prefabricated integrated circuitry. First, a photoresist sacrificial layer is spun on the wafer and patterned to

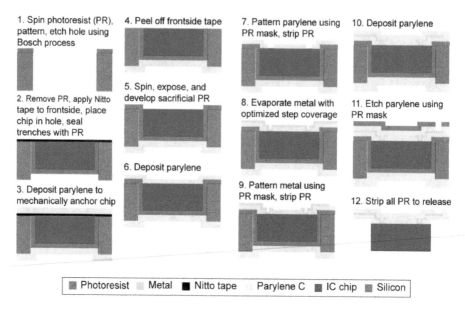

1. Spin photoresist (PR), pattern, etch hole using Bosch process

2. Remove PR, apply Nitto tape to frontside, place chip in hole, seal trenches with PR

3. Deposit parylene to mechanically anchor chip

4. Peel off frontside tape

5. Spin, expose, and develop sacrificial PR

6. Deposit parylene

7. Pattern parylene using PR mask, strip PR

8. Evaporate metal with optimized step coverage

9. Pattern metal using PR mask, strip PR

10. Deposit parylene

11. Etch parylene using PR mask

12. Strip all PR to release

⬛ Photoresist ░ Metal ⬛ Nitto tape ░ Parylene C ⬛ IC chip ⬛ Silicon

Figure 28. Detailed process flow for CL-I² package fabrication.

expose the chip's surface. After baking to remove excess solvent, approximately 3 μm of parylene C is deposited on the entire wafer. Photoresist is spun on the wafer, exposed in the 10X reduction stepper, and developed to pattern etch holes above the on-chip pads, similar to the vias in the dual metal-layer process. This pattern is transferred into the parylene using an O_2 plasma in an RIE system, exposing the metal of these on-chip pads. Two hundred angstroms of titanium and 2000 Å of gold are deposited in the e-beam evaporation system using optimized step coverage, and patterned (using a photoresist mask and wet etching) to define the remote contact pads and remote pad to on-chip pad interconnects. The top photoresist is stripped, and a second layer of approximately 10 μm of parylene C is deposited and patterned as before, but this time to open the remote pads/electrodes to enable electrical testing. Finally, all photoresist, including the sacrificial layer, is removed by soaking the wafer in acetone, releasing a flexible parylene skin with embedded interconnects to the packaged ASIC. The host wafer can be substituted in the process with a precisely machined substrate, and can be reused after this release step. It is also important to note that the ASIC or discrete component can be of any thickness, but generally the thickness should be less than that of the host wafer or machined substrate, and it can have parylene or any hermetic coating deposited on it *a priori*, provided that the chip contacts can be opened using microfabrication techniques before the interconnect metal is laid down and patterned (Figure 28 steps 8 and 9). Thus, this technology combines the best aspects of chip-level packaging, in which every surface of the prefabricated chip can be manipulated or coated beforehand, and wafer-level packaging, in that photolithography and microfabrication can be performed on the surface after such chip-level techniques.

9.3. Integration testing results

We successfully performed photolithography on ten prefabricated stand-alone chips using this paradigm: seven test chips (three conformally coated in parylene *a priori*) and three MOSIS-fabricated chips (one coated in parylene *a priori*). A MOSIS chip anchored in place in the host substrate is shown centrally in Figure 29 with the host wafer shown on the perimeter. As is expected, minimizing vertical displacement of the ASIC from the wafer surface is crucial for further photolithography steps. Figure 30 gives typical surface profiles of all ten chips with respect to the surrounding host wafer, and indicates that for most chips, this vertical displacement was less than 5 μm after removal of the frontside Nitto tape. Photolithography on the somewhat anomalous test chips 2 and 7, however, was also successful.

The accurate horizontal alignment of the perimeter interconnects to the embedded chips is shown in Figure 31, with Figure 32 giving a detailed micrograph of a single interconnect for both the test chip (a) and the MOSIS chip (b). By design, the chips should be self-aligned to within 10 μm of lateral displacement; some chips were aligned far better than this, however others were misaligned worse than this. With tighter tolerances on the cavity sidewalls, or with chip-alignment lithographic equipment, this alignment error could be improved. The embedded chip with remote contact pads is shown in Figure 33, and Figure 34 depicts the flexibility of this package. Functional contacts to the chips were verified as described in [40].

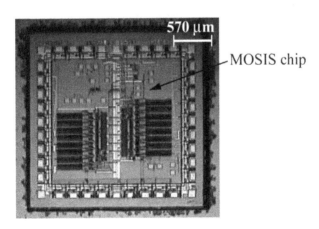

Figure 29. MOSIS chip (center) shown anchored in host silicon substrate (perimeter).

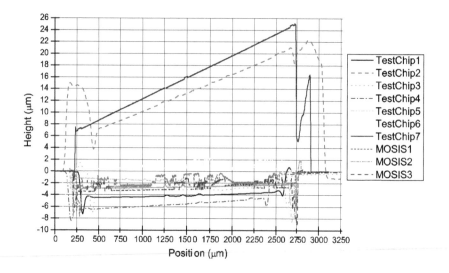

Figure 30. Typical single-axis vertical displacements of all 10 chips after mechanical anchoring in the host wafer (Figure 28, step 4), where the top surface of the wafer corresponds to 0 μm.

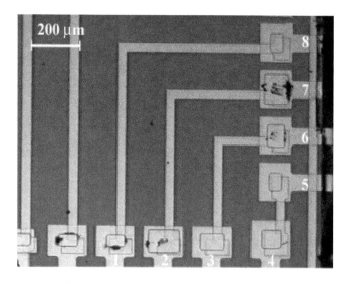

Figure 31. Embedded chip with fabricated perimeter interconnects (numbered traces connect to numbered remote pads shown in Figure 33).

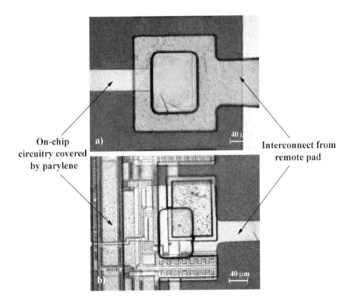

Figure 32. (a) Example of <10 μm lateral misalignment of a test chip; (b) Example of >10 μm lateral misalignment of a MOSIS chip.

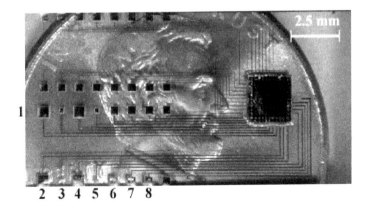

Figure 33. A CL-I² packaged chip shown resting on a penny.

Figure 34. Demonstrates flexibility of CL-I² package. Chip can be seen underlying overlying parylene "skin."

9.4. Discussion

It should be stressed that the lead-count and interconnect density limitations for this technology stem only from the limitations of the microfabrication and photolithography equipment used to fabricate the CL-I² package, and, in particular, to pattern the first parylene etch (Figure 28, step 7). All interconnects to the chip are fabricated simultaneously during the metal deposition step, and depend on optimal step coverage of the parylene sidewall (aided in part by the slightly isotropic nature of the O_2 plasma etch of parylene [47]). The CL-I² process thus avoids the use of tedious and comparatively low-density ball-wedge [38] or wire bonding.

Our method of incorporating discrete modules into a MEMS process is far more cost-effective when compared with full-wafer IC processing and MEMS integration [51], because valuable space on the wafer is not wasted during the IC fabrication step. Furthermore, in comparison to other ASIC integration attempts [51-53], this packaging scheme is superior for biodevices because it takes advantage of parylene's low water-absorption [54] and highly conformal pinhole-free deposition, and because the package is both flexible and biocompatible. Among the feasible uses for this technology is the interconnection of chips, devices such as other CMOS-compatible MEMS, as well as discrete components such as chip capacitors, fabricated using different materials and processes, to make large conglomerate circuits for neural prostheses and for other applications. This technology is capable of increasing the previously projected number of I/O interconnects available in 2010 significantly, while using lead-free, biocompatible materials. Fabrication is not limited to the use of parylene as either the backside anchoring material or as the frontside electrode insulation material, although, because of its superior electrical, mechanical, and water permeability properties when compared with other polymers, we believe parylene will ultimately prove to be the best choice for monolithic high-density neural prosthetics. It is interesting to note that another research group has, after our original publications [40, 56, 57], explored an integration technique in polyimide very similar to ours, with interconnect density motivations much akin to our own [58]. We have recently demonstrated a fully integrated parylene-based single-channel neural stimulator [59]. *In vitro* measurements demonstrated the ability to generate 7 V pulses of 500 μs pulse width from a wireless transmitter 4 mm away. Preliminary evidence from animal implantation studies has shown these are mechanically reliable under surgical conditions.

10. Conclusions and future work

BioMEMS, an emerging field in which MEMS are designed, fabricated, and utilized to interface with and examine biological systems, is a discipline replete with incredible possibilities, but also one that is fraught with potential pitfalls. This is no more true than in the case of implantable microelectronics, where applications abound because of the near perfect match in the sizes of the functional components of the body, namely cells and neurons, and the technologies possible using microfabrication techniques. Materials must be carefully selected such that they are biocompatible with the body, while still enabling maximal functionality to be delivered to the patient, requirements that are often competing in nature. In this vein, the microtechnologies necessary for parylene-based flexible microelectrode technologies have been presented. All evidence thus far points to the fact that such parylene-based technologies are likely an ideal option for implantable neuroprostheses and microdevices.

It should be stressed that these technologies are not limited to use in retinal and spinal cord prostheses. In fact, the ability of our flexible arrays to conform to the geometries of interest in the human body enables them to be used in a variety of locations heretofore previously inaccessible with such high precision. Such locations include the surfaces of the cerebral cortex, another area of interest from both a scientific and treatment point of view due to such neurological problems as stroke, epilepsy, and memory loss. Areas of other interest include peripheral nerve and muscle. It is also possible to embed sensors in such arrays, as has recently been demonstrated [60], to assess mechanical forces placed on the tissues of interest by our arrays as well as to detect extrinsic pressures, such as those within the eye or within blood vessels.

Understandably, there is a public reticence to the implementation of such technologies in the human body. While such apprehension is not a recent phenomenon, the burgeoning era of computerized special effects in television and cinema has helped fuel the fear that the blending of "man" and "machine" can have devastating consequences. What is missed in such intimations is that, in the hands of ethical doctors, engineers, and other scientists, such consequences are extremely remote. But it is not about the inventors of these technologies, and it should never be about personal glory. All of that slips away the moment one talks to a person who has devoted their life to be a pioneer in the field by volunteering to be a test subject of such devices for the benefit of mankind. In such conversations, one realizes the full potential of this technology. Investigation into these devices not only has the possibility to positively affect the lives of such people, by enabling them to "walk" or "see" again, but it transcends all that by bringing about in all involved a sense of camaraderie. Indeed, the selfless motivation of such individuals who devote their most precious commodity, their body, to such studies, more than anything else highlights our very humanity. It shows that, despite the need for such technological innovation, it is only by working together and for one another that we, as a people, can break the bonds of human disease.

Author details

Damien C. Rodger[1], Wen Li[2], James D. Weiland[1], Mark S. Humayun[1] and Yu-Chong Tai[3]

1 University of Southern California, Los Angeles, CA, USA

2 Michigan State University, East Lansing, MI, USA

3 California Institute of Technology, Pasadena, CA, USA

References

[1] The Eye Diseases Prevalence Research Group, "Prevalence of age-related macular degeneration in the United States," *Arch Ophthalmol,* vol. 122, pp. 564-572, April 1 2004.

[2] G. Fishman, V. Vasquez, M. Fishman, and D. Berger, "Visual loss and foveal lesions in Usher's syndrome," *Br J Ophthalmol,* vol. 63, pp. 484-488, Jul 1 1979.

[3] B. W. Jones, C. B. Watt, J. M. Frederick, W. Baehr, C.-K. Chen, E. M. Levine, A. H. Milam, M. M. Lavail, and R. E. Marc, "Retinal remodeling triggered by photoreceptor degenerations," *The Journal of Comparative Neurology,* vol. 464, pp. 1-16, 2003.

[4] E. Margalit and S. R. Sadda, "Retinal and Optic Nerve Diseases," *Artificial Organs,* vol. 27, pp. 963-974, 2003.

[5] A. Santos, M. S. Humayun, E. de Juan, Jr., R. J. Greenburg, M. J. Marsh, I. B. Klock, and A. H. Milam, "Preservation of the inner retina in retinitis pigmentosa. A morphometric analysis," *Arch Ophthalmol,* vol. 115, pp. 511-515, April 1, 1997 1997.

[6] R. E. MacLaren, R. A. Pearson, A. MacNeil, R. H. Douglas, T. E. Salt, M. Akimoto, A. Swaroop, J. C. Sowden, and R. R. Ali, "Retinal repair by transplantation of photoreceptor precursors," *Nature,* vol. 444, pp. 203-207, 2006.

[7] M. Humayun, R. Propst, E. de Juan, Jr., K. McCormick, and D. Hickingbotham, "Bipolar surface electrical stimulation of the vertebrate retina," *Arch Ophthalmol,* vol. 112, pp. 110-116, Jan 1 1994.

[8] M. S. Humayun, E. de Juan, Jr., G. Dagnelie, R. J. Greenberg, R. H. Propst, and D. H. Phillips, "Visual perception elicited by electrical stimulation of retina in blind humans," *Arch Ophthalmol*, vol. 114, pp. 40-46, January 1, 1996 1996.

[9] M. A. Schiefer and W. M. Grill, "Sites of Neuronal Excitation by Epiretinal Electrical Stimulation," *IEEE Transactions on Neural Systems and Rehabilitation Engineering*, vol. 14, pp. 5-13, 2006.

[10] A. Y. Chow and V. Y. Chow, "Subretinal electrical stimulation of the rabbit retina," *Neuroscience Letters*, vol. 225, pp. 13-16, Mar 28 1997.

[11] C. Veraart, M. C. Wanet-Defalque, B. Gerard, A. Vanlierde, and J. Delbeke, "Pattern recognition with the optic nerve visual prosthesis," *Artificial Organs*, vol. 27, pp. 996-1004, Nov 2003.

[12] E. M. Schmidt, M. J. Bak, F. T. Hambrecht, C. V. Kufta, D. K. ORourke, and P. Vallabhanath, "Feasibility of a visual prosthesis for the blind based on intracortical microstimulation of the visual cortex," *Brain*, vol. 119, pp. 507-522, Apr 1996.

[13] E. M. Maynard, "Visual prostheses," *Annual Review of Biomedical Engineering*, vol. 3, pp. 145-168, 2001.

[14] J. S. Pezaris and R. C. Reid, "Demonstration of artificial visual percepts generated through thalamic microstimulation," *PNAS*, vol. 104, pp. 7670-7675, May 1, 2007 2007.

[15] M. S. Humayun, J. D. Weiland, G. Y. Fujii, R. Greenberg, R. Williamson, J. Little, B. Mech, V. Cimmarusti, G. Van Boemel, and G. Dagnelie, "Visual perception in a blind subject with a chronic microelectronic retinal prosthesis," *Vision Research*, vol. 43, pp. 2573-2581, Nov. 2003.

[16] M. S. Humayun, R. Freda, I. Fine, A. Roy, G. Fujii, R. J. Greenberg, J. Little, B. Mech, J. D. Weiland, and E. de Juan, "Implanted intraocular retinal prosthesis in six blind subjects," *Investigative Ophthalmology & Visual Science*, vol. 46, p. 1144, 2005.

[17] B. S. Wilson, C. C. Finley, D. T. Lawson, R. D. Wolford, D. K. Eddington, and W. M. Rabinowitz, "Better speech recognition with cochlear implants," *Nature*, vol. 352, pp. 236-8, Jul 18 1991.

[18] J. D. Weiland, W. Liu, and M. S. Humayun, "Retinal prosthesis," *Annual Review of Biomedical Engineering*, vol. 7, pp. 361-401, 2005.

[19] C. J. Lee, J. A. Vroom, H. A. Fishman, and S. F. Bent, "Determination of human lens capsule permeability and its feasibility as a replacement for Bruch's membrane," *Biomaterials*, vol. 27, pp. 1670-8, Mar 2006.

[20] L. J. Noble and J. R. Wrathall, "Correlative Analyses of Lesion Development and Functional Status after Graded Spinal-Cord Contusive Injuries in the Rat," *Experimental Neurology*, vol. 103, pp. 34-40, Jan 1989.

[21] K. Kurohane, A. Tominaga, K. Sato, J. R. North, Y. Namba, and N. Oku, "Photodynamic therapy targeted to tumor-induced angiogenic vessels.," *Cancer Lett*, vol. 167, pp. 49-56, Jun 2001.

[22] K. E. Krishnan L, Jewell WR, "Immediate effect of irradiation on microvasculature.," *Int J Radiat Oncol Biol Phys*, vol. 15, pp. 1147-1450, Jul 1988.

[23] J. W. Little, J. F. Ditunno, S. A. Stiens, and R. M. Harris, "Incomplete spinal cord injury: Neuronal mechanisms of motor recovery and hyperreflexia," *Archives of Physical Medicine and Rehabilitation*, vol. 80, pp. 587-599, May 1999.

[24] J. H. Burridge, P. N. Taylor, S. A. Hagan, D. E. Wood, and I. D. Swain, "The effects of common peroneal stimulation on the effort and speed of walking: a randomized controlled trial with chronic hemiplegic patients," *Clinical Rehabilitation*, vol. 11, pp. 201-210, Aug 1997.

[25] J. Rozman, R. Acimovicjanezic, I. Tekavcic, M. Kljajic, and M. Trlep, "Implantable Stimulator for Selective Stimulation of the Common Peroneal Nerve - a Preliminary-Report," *Journal of Medical Engineering & Technology*, vol. 18, pp. 47-53, Mar-Apr 1994.

[26] P. Taylor, J. Burridge, A. Dunkerley, D. Wood, J. Norton, C. Singleton, and I. Swain, "Clinical audit of 5 years provision of the Odstock dropped foot stimulator," *Artificial Organs*, vol. 23, pp. 440-442, May 1999.

[27] K. W. Horch and G. S. Dhillon, *Neuroprosthetics theory and practice*. River Edge, N.J.: World Scientific, 2004.

[28] R. Brissot, P. Gallien, M. P. Le Bot, A. Beaubras, D. Laisne, J. Beillot, and J. Dassonville, "Clinical experience with functional electrical stimulation-assisted gait with parastep in spinal cord-injured patients," *Spine*, vol. 25, pp. 501-508, Feb 15 2000.

[29] D. Graupe and K. H. Kohn, "Functional neuromuscular stimulator for short-distance ambulation by certain thoracic-level spinal-cord-injured paraplegics," *Surgical Neurology*, vol. 50, pp. 202-207, Sep 1998.

[30] E. Roederer, N. H. Goldberg, and M. J. Cohen, "Modification of retrograde degeneration in transected spinal axons of the lamprey by applied DC current," *J. Neurosci.*, vol. 3, pp. 153-160, January 1, 1983 1983.

[31] C. C. Stichel and H. W. Muller, "Experimental strategies to promote axonal regeneration after traumatic central nervous system injury," *Progress in Neurobiology*, vol. 56, pp. 119-148, 1998.

[32] S. A. Maguire AM, *Radiation retinopathy. In: Retina*. St. Louis: Mosby-Year book., 1994.

[33] M. Szwarc, "Some Remarks on the CH2=Benzene=CH2 Molecule," *Discussions of the Faraday Society*, vol. 2, pp. 46-49, 1947.

[34] W. F. Gorham, "A New General Synthetic Method for Preparation of Linear Poly-P-Xylylenes," *Journal of Polymer Science Part A-1: Polymer Chemistry,* vol. 4, pp. 3027-3039, 1966.

[35] L. Wolgemuth, "Crystal-Clear Coating Covers Components," *Medical Design,* vol. 6, pp. 48-51, December 1 2006.

[36] "New SCS Parylene HT Conformal Coating Ideal For High-Temperature Applications." vol. 2007: Specialty Coating Systems.

[37] "Reliable protection for advanced electronics." vol. 2007: Specialty Coating Systems.

[38] T. Stieglitz, W. Haberer, C. Lau, and M. Goertz, "Development of an inductively coupled epiretinal vision prosthesis," in *Int. IEEE Eng. in Med. and Biol. Soc. Meet.,* San Francisco, CA, USA, 2004, pp. 4178-4181 Vol.6.

[39] A. C. Hoogerwerf and K. D. Wise, "A three-dimensional microelectrode array for chronic neural recording," *IEEE Transactions on Biomedical Engineering,* vol. 41, pp. 1136-1146, 1994.

[40] D. C. Rodger, J. D. Weiland, M. S. Humayun, and Y.-C. Tai, "Scalable high lead-count parylene package for retinal prostheses," *Sensors and Actuators B: Chemical,* vol. 117, pp. 107-114, 2006.

[41] W. Li, D. C. Rodger, E. Meng, J. D. Weiland, M. S. Humayun, and Y.-C. Tai, "Flexible Parylene Packaged Intraocular Coil for Retinal Prostheses," in *Int. IEEE-EMBS Microtech. in Med. and Bio. Meet.,* Okinawa, Japan, 2006, pp. 105-108.

[42] P. J. Chen, D. C. Rodger, M. S. Humayun, and Y. C. Tai, "Unpowered spiral-tube parylene pressure sensor for intraocular pressure sensing," *Sensors and Actuators a-Physical,* vol. 127, pp. 276-282, Mar 13 2006.

[43] P. J. Chen, D. C. Rodger, E. M. Meng, M. S. Humayun, and Y. C. Tai, "Surface-micromachined parylene dual valves for on-chip unpowered microflow regulation," *Journal of Microelectromechanical Systems,* vol. 16, pp. 223-231, Apr 2007.

[44] E. Meng, P.-J. Chen, D. Rodger, Y.-C. Tai, and M. Humayun, "Implantable parylene MEMS for glaucoma therapy," in *Int. IEEE-EMBS Microtech. in Med. and Bio. Meet.,* Oahu, HI, USA, 2005, pp. 116-119.

[45] A. K. Ahuja, M. R. Behrend, M. Kuroda, M. S. Humayun, and J. D. Weiland, "Spatial Response Properties of Electrically Stimulated Retina," *Invest. Ophthalmol. Vis. Sci.,* vol. 48, p. 4444, May 10 2007.

[46] A. P. Chandran, K. Oda, H. Shibasaki, and M. Pisharodi, "Spinal somatosensory evoked potentials in mice and their developmental changes," *Brain Dev,* vol. 16, pp. 44-51, Jan-Feb 1994.

[47] E. Meng and Y.-C. Tai, "Parylene etching techniques for microfluidics and bio-MEMS," in *IEEE Int. Conf. on MEMS,* Miami, FL, USA, 2005, pp. 568-571.

[48] C. A. Curcio and K. A. Allen, "Topography of Ganglion-Cells in Human Retina,"
      Journal of Comparative Neurology, vol. 300, pp. 5-25, Oct 1990.

[49] J. J. Whalen, J. D. Weiland, and P. C. Searson, "Electrochemical Deposition of Plati-
      num from Aqueous Ammonium Hexachloroplatinate Solution," *Journal of The Electro-
      chemical Society*, vol. 152, pp. C738-C743, 2005.

[50] S. B. Brummer and M. J. Turner, "Electrical Stimulation with Pt Electrodes: I-A Meth-
      od for Determination of "real" electrode Areas," *IEEE Transactions on Biomedical Engi-
      neering*, vol. 24, pp. 436-439, 1977.

[51] Y. Xu, Y.-C. Tai, A. Huang, and C.-M. Ho, "IC-integrated flexible shear-stress sensor
      skin," *Microelectromechanical Systems, Journal of*, vol. 12, pp. 740-747, 2003.

[52] J. T. Butler, V. M. Bright, and J. H. Comtois, "Advanced multichip module packaging
      of microelectromechanical systems," in *Int. Conf. on Solid-State Sensors and Actuators*,
      Chicago, IL, USA, 1997, pp. 261-264.

[53] W. Daum, W. E. Burdick, and R. A. Fillion, "Overlay high-density interconnect: a
      chips-first multichip module technology," *Computer*, vol. 26, pp. 23-29, 1993.

[54] J. J. Licari and L. A. Hughes, *Handbook of polymer coatings for electronics : chemistry,
      technology, and applications*, 2nd ed. Park Ridge, N.J., U.S.A.: Noyes Publications, 1990.

[55] R. R. Tummala, *Fundamentals of microsystems packaging*. New York: McGraw-Hill,
      2001.

[56] D. C. Rodger and Y. C. Tai, "Microelectronic packaging for retinal prostheses," *IEEE
      Engineering in Medicine and Biology Magazine*, vol. 24, pp. 52-57, Sep-Oct 2005.

[57] D. C. Rodger, J. D. Weiland, M. S. Humayun, and Y.-C. Tai, "Scalable flexible chip-
      level parylene package for high lead count retinal prostheses," in *Int. Conf. Solid-State
      Sensors, Actuators and Microsystems*, Seoul, Korea, 2005, pp. 1973-1976.

[58] H. Sharifi, T. Y. Choi, and S. Mohammadi, "Self-aligned wafer-level integration tech-
      nology with high-density interconnects and embedded passives," *IEEE Transactions
      on Advanced Packaging*, vol. 30, pp. 11-18, Feb 2007.

[59] W. Li, D. C. Rodger, A. Pinto, E. Meng, J. D. Weiland, M. S. Humayun, and Y.-C. Tai,
      "Parylene-based integrated wireless single-channel neurostimulator," *Sensors and Ac-
      tuators A: Physical*, vol. 166, pp. 193-200.

[60] B. J. Kim, C. A. Gutierrez, G. A. Gerhardt, and E. Meng, "Parylene-based electro-
      chemical-MEMS force sensor array for assessing neural probe insertion mechanics,"
      in *Micro Electro Mechanical Systems (MEMS), 2012 IEEE 25th International Conference
      on*, pp. 124-127.

Chapter 4

Optical MEMS for Telecommunications: Some Reliability Issues

Ivanka Stanimirović and Zdravko Stanimirović

Additional information is available at the end of the chapter

1. Introduction

Constant demand for mobility, interconnectivity and bandwidth is causing rapid expansion of the telecommunication infrastructure across the world. World-wide installation of optical fibre-based telecommunication systems has given rise to a promising optically-related sub segment of MEMS technology called micro-opto-electro-mechanical systems (MOEMS), commonly known as optical MEMS. MEMS telecommunications applications can be roughly divided into two key classes: optoelectronic packaging and functional optical devices. When functional optical devices are in question, optical MEMS devices that integrate optical, mechanical, and electrical components on a single wafer are allowing the implementation of various key optical-network elements in a compact, low-cost form. They usually involve small moving optical parts in order to obtain more advanced functionality. In optoelectronic packaging, MEMS are providing low-cost accurate optical alignment. At the moment, fabrication of complex optical MEMS devices and micro-electro-mechanical alignment devices is based on micromachining techniques combined with IC-based processing methods. Such manufacturing techniques have enabled low cost, mass production of optical MEMS components and devices. However, successful commercialization of optical MEMS technology that is being driven by the progress in optical communications strongly depends on device reliability. Optical MEMS device reliability is significantly more complex than silicon IC reliability, partly because optical MEMS failures can be either electrical or mechanical, and partly because there is a vast diversity of device designs, materials and functions. It is of the greatest importance that design and realization of optical MEMS device must include all levels of reliability issues from the onset of the project. For that reason, this chapter focuses on the identification and understanding of main mechanisms that cause failure of optical MEMS devices that are being used in telecommunications. First, the commonly used MEMS processing technologies are summarized. Then, functional optical MEMS devices for optical network infrastructure are discussed. Finally, the key issues of various MEMS device failure mechanisms and design, processing and packaging implications are presented. At the closing subsection, the brief summary of the topic is presented with an emphasis on the importance of the research of relevant reliability issues that stand in the way of successful commercialization of optical MEMS devices.

2. Optical MEMS technologies

Similar to optical MEMS devices, there is no single standard processing technology for optical MEMS fabrication. Silicon based optical MEMS is dominant materials system and different micromachining processes are being used as the most appropriate fabrication techniques. Also, conventional IC processes (lithography, depositions, implantation, dry etching, etc.) are often used in microstructure formation.

Bulk micromachining has been used for a long time for realization of 3D optomechanical structures on Si substrate for aligning optical fibres or forming optical MEMS devices. Single crystal Si has excellent mechanical properties and low-cost, high-purity Si substrates are available from IC industry. Si bulk micromachining is the process that impacts the substrate. Precise removal of the designated part of silicon substrate can be achieved by anisotropic etchants. Large difference in anisotropic etch rates between the <111> plane and other crystallographic planes in Si, enables pattern formation on either front-side or backside of the substrate. The etching rate of anisotropic etchants such as potassium hydroxide (KOH), aqueous solution of ethylene diamine and pyrocatechol (EDP) and tetramethylammonium hydroxide (TMAH), is much slower in <111> direction than in <100> and <110> directions [1]. Selectivity for such anisotropic etchants can be higher than 100 allowing creation of 3D optomechanical structures with high precision. Basic properties of commonly used anisotropic etchants are listed in Table 1.

Etchant	Etch rate (110) μm/min	AR {100}/{111}	Etch Masks	Etch stop	Main characteristics
KOH	1,4	400	Si_3N_4, SiO_2	$B>10^{20}/cm^3$	Fastest, greatest selectivity, makes vertical sidewalls
EDP	1,25	35	SiO_2, Si_3N_4, Ta, Au, Cr, Ag, Cu	$B\sim7\times10^{19}/cm^3$	Lots of masks, lowest Boron doping etch stop, low AR
TMAH	1	30	Si_3N_4	$B\sim4\times10^{20}/cm^3$	Smooth surface, slow etch rate, low AR

Table 1. Basic properties of common anisotropic etchants

V-shaped grooves commonly used for precision positioning of optic fibres, are an example of this processing technology. The (100) Si substrate is first masked with an etch-resistance surface layer (deposited Si_3N_4 for KOH or thermally grown SiO_2 for EDP) and then the Si is etched. Slower rate of <111> planes etch enables V-groove formation by etching <100> oriented planes. V-groove depth can be very well controlled by lithography because {111} planes are effective stop etching planes. Schematic of V-shaped groove formation is shown in Figure 1. By etching through square openings, pyramidal-shaped holes can also be formed that are being used for holding ball lenses. V-grooves and pyramidal-shaped holes are the basis of conventional microoptical benches. Bulk optical components are placed on the etched Si substrate and precisely positioned by holes of various geometries. Vertical micromirrors can be formed by anisotropic etching on a (110) Si substrate. Atomically smooth {111} planes are perpendicular to the surface of the substrate. Large-area semitransparent, optical-quality surfaces are provided. These micromirrors can be also used as beam splitters. In addition to the {111} stop etch planes, some etchants exhibit reduced etch rate in regions that are heavily doped with boron. This allows more flexibility in shapes of final structures: membranes, suspended beams, support beams for vertical micromirrors etched on (110) substrate, etc. Besides boron, other doping materials can be used but doping involves high temperatures and has side effects such as lattice shrinkage and introduction of large tensile stresses in parts formed this way [2].

Figure 1. V-shaped grooves formed by bulk silicon etch with wet chemistry

Fusion bonding of glass to bulk micromachined Si substrates allows formation of encapsulated structures as shown in Figure 2. Also, multilayer structures may be formed by bonding Si substrates together. In this way, the range of devices that can be manufactured using bulk micromachining is greatly extended.

Figure 2. Wafer bonding

Another process commonly used in optical MEMS fabrication is surface micromachining. While, in bulk micromachining, substrates materials are being removed in order to create 3D structures, surface micromachined structures are constructed from deposited thin films. Alternating layers of structural and sacrificial materials are deposited and patterned on the substrate. The sacrificial layers can be selectively removed by an etchant that attacks only the sacrificial materials. In this way suspended beams, cantilevers, diaphragms and cavities can be realized. Because of its excellent mechanical properties, polysilicon is being used as structural material and SiO_2 as the sacrificial material because of the high selectivity of sacrificial etching with hydrofluoric acid. Figure 3. illustrates polysilicon surface machining process. The complexity of the surface machining process is determined by the number of structural and sacrificial layers. Two structural layers allow formation of free moving mechanical gears, springs, sliders, etc. The main advantage of surface micromachining over bulk micromachining is that many different devices can be realized using common fabrication process. By changing patterns on the photomask layouts different devices are being fabricated simultaneously on the same substrate. For that reason, the surface micromachining process is often referred to as an IC process that allows formation of multilayer structures usually with two to five polysilicon levels.

Figure 3. Polysilicon surface micromachining

Often, it is desirable to fabricate structures thicker than those achievable using polysilicon. An alternative micromachining process uses lithographic exposure of thick photoresist, followed by electroplating to build on chip high aspect ratio 3D structures. In the LIGA (lithography, electroplating and moulding) process synchotron radiation is used as the exposure source that can achieve feature heights of the order of 500μm. Cheaper alternatives use excimer lasers or UV mask aligners that achieve feature heights of the order of 200μm and 20μm, respectively [2]. Parts are usually plated in nickel after removal of the resist as illustrated in Figure 4. The released metal layer can be used in various applications including optical MEMS devices.

Figure 4. Metal micromachining

Suspended single crystal Si structures, with lower stress and more reproducible properties than polysilicon, are formed using process based on BSOI (bonded silicon-on-insulator). Si wafer is thermally bonded to an oxidized Si substrate. Desired thickness (usually 5 to 200μm [3]) of the bonded wafer is achieved by polishing and the bonded layer is structured by deep reactive ion etching (DRIE) that has high etch rates and anisotropy to form very deep features with almost vertical sidewalls (Figure 5.). Movable parts can be made by removal of the buried oxide and one of the typical applications of this technique is realization of vertical mirrors for optical switching.

Figure 5. Deep reactive ion etching (DRIE) of bonded silicon-on-insulator (BSOI)

DRIE has also allowed Si micromolding techniques, such as HexSil process, to be developed [4]. DRIE is used to etch narrow trenches into the substrate. Trenches are fraction of a millimetre deep. After that, a sacrificial oxide layer is deposited, followed by the polysilicon structural layer that fills the trenches. As shown in Figure 6., deep suspended structures are being made by releasing the polysilicon.

Figure 6. HexSil process

All described techniques involve surface patterning processes and therefore realized microstructures are quasi 3D. Very often out-of-plane structures with high aspect ratios are required for free-space optical systems. Anisotropic etching or deep dry etching can provide such structures but it is difficult to pattern their side walls. Fully 3D structures can be formed using microhinge technology [5]. Surface micromachined polysilicon planes are patterned by photolithography and then folded into 3D structures. Figure 7. shows schematic cross section of the microhinge that consists of hinge pin and a confining staple. After selective etching of the sacrificial SiO$_2$, the polysilicon plate connected to the hinge pin is free to rotate out of the substrate plane and become perpendicular to the substrate. Polysilicon plate can also achieve other angles. This technology allows monolithic integration of 3D structures with surface micromachined actuators. It is of the special interest for fabrication of integrable free-space microoptical elements.

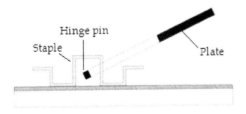

Figure 7. Schematic of surface-micromachined microhinge

The full potential of surface micromachining, bulk micromachining and wafer/chip bonding techniques is still being explored. The key activities are continued development of masks and etches that can yield high aspect ratio structures and the development of deposition techniques. Special attention is being paid to the development of techniques for creating fully 3D structures.

3. Optical MEMS devices

Components fabricated using optical MEMS devices are finding an increasing number of applications when optical side of telecommunications is in question. They can be divided in two categories: core and peripheral optical MEMS devices (Table 2.). Core optical MEMS devices incorporate fixed structures (V-groves, gratings, etc.) and moving elements (micromirrors, attenuators, etc.). Peripheral optical MEMS devices are alignment components and structural components. The key area, when optical MEMS for telecommunications are in question, is related to functional optical devices - devices that involve small moving optical parts necessary for more advanced functionality. They are core optical MEMS with moving elements.

Core optical MEMS		Peripheral optical MEMS	
Fixed Structures	V-grooves	Alignment components	Lenses
	Connectors		
	Benches		
	Gratings		
Moving Elements	Mirrors	Structural components	Packaging
	Shutters		Beam steering
	Filters		Fiber-guides
	Attenuators		

Table 2. Optical MEMS for telecommunication applications [6]

One of the simplest functional optical MEMS devices is the variable optical attenuator (VOA) [2, 7]. Typically, a moving micro-structure is designed to either partially block or decouple the lightpath. An example of a blocking VOA is shown in Figure 8. The light from the input fiber is collimated with a lens, partially blocked or attenuated by the MEMS device and recoupled to an output fiber. The MEMS device itself could be actuated horizontally or vertically. Actuators could be electrostatic, thermal or electromagnetic. Such a device could be also used as an on-off switch.

The main goal of optical MEMS is providing a high-performance, low-cost solutions for optical switching and wavelength division multiplexing (WDM) or dense wavelength division multiplexing (DWDM). Depending on the specific application, these devices can be wavelength insensitive or wavelength selectable. Wavelength and protocol insensitive device for

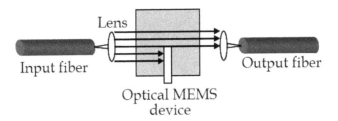

Figure 8. Schematic of variable optical attenuator (VOA)

all optical switching is optical cross connect (OXC). It replaced conventional optical-electrical-optical (OEO) switching that required conversion of optical signals to electrical ones, switching of electrical signals and conversion of electrical signals to optical ones. OEO switching solution cannot keep up with rapid data rate increase because expensive transceivers and electrical switch core will have to be replaced. However, all optical switching provide avoidance of conversion stages and core switch is independent of data rate and data protocol, making cross connect ready for data rate upgrades. This solution is also cost effective because the use of expensive power-consuming high-speed electronics, transmitters and receivers is avoided. This complexity reduction significantly improves reliability of the device. A typical MEMS OXC consists of micromirrors made of either polysilicon or crystalline silicon, using silicon-on-oxide (SOI), coated with metal for reflectivity. The actuation can be electrostatic, magnetic or combination of the two. Two MEMS approaches for optical switching can be distinguished: 2D (planar) switching and 3D free-space switching [8, 9]. In 2D MEMS the switches are digital because mirror position is bistable (Figure 9.). MEMS micromirrors are arranged in a crossbar configuration and all optical paths lie on a planar (2D) surface (Figure 10.). When a micromirror is activated it moves into the path of the beam and directs the light to one of the outputs. Light can also be passed through the matrix without hitting the micromirror allowing adding or dropping optical channels.

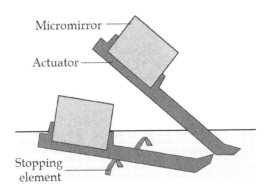

Figure 9. Schematic of basic element for 2D optical switches

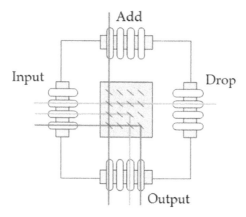

Figure 10. MEMS approach for optical cross connect switching

For switching ultra-high N networks planar switching is being replaced with more robust and cost-effective solution. 3D MEMS is a most promising technology for optical cross connect switches with >1000 input and output ports. In 3D MEMS a connection path is established by tilting two micromirrors independently to direct the light from an input port to selected output port (Figure 11.). This approach requires 2-axis mirror cells that usually consist of a gimbal and a mirror [10]. The gimbal connects to the support structure with a pair of torsional springs and another pair of torsional springs connects the mirror to the gimbal. Second pair of springs is rotated 90° with the respect to the first pair. Each pair of springs san be independently actuated and their combination enables two-directional tilt of the mirror (Figure 12.). A drawback of this approach is that a complex and expensive feedback system is required to maintain the position of the mirrors during external disturbances or drift.

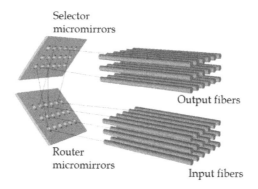

Figure 11. 3D MEMS approach for optical cross connect switching

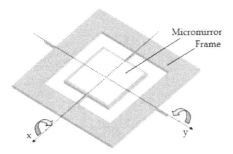

Figure 12. Schematic of two-axes single crystal silicon mirror

The output characteristics of an optical amplifier are not uniform across the laser wavelength spectrum. This is problematic for WDM because each segment of spectrum carries a data channel. For that reason, a dynamic gain equalizer (DGE) is needed to level output spectrum [10]. First, channels are separated by dispersing spectrum through assembly of lens and grating (Figure 13.). Then, they are projected onto the DGE and the output of each channel is tuned independently. The tuning can be performed by either an MEMS micromirror array (Figure 14.) or mechanical anti-reflection switch (MARS) (Figure 15.). MARS uses the strip of dielectric (usually silicon-nitride) membrane and air gap that serve as a spatially variable, tunable multi-layer dielectric mirror. When the incident signal is spectrally dispersed along the axis of the device defined by an array of strip electrodes, one obtains a simple and compact tunable spectral shaper.

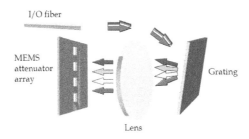

Figure 13. Schematic of WDM DGE using MEMS mirror array, lens and grating

In optical WDM provisioning, a data channel may be dropped or added. MEMS technology provides simple solution in optical add-drop multiplexer (OADM) [10]. The tilting-mirror DGE at larger tilting angles can achieve dynamic channel blocking. The full add-drop function of data channels is obtained by inserting channel blocker between an optical splitter and an optical coupler. Independent control for each cannel is wavelength selectable resulting in flexibility to add/drop any combination of data channels. WDM provisioning using wave-

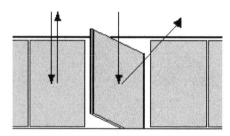

Figure 14. Tilting micromirrors - schematic of operation

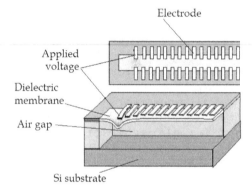

Figure 15. Schematic of MARS DGE using continuous membrane and finger electrodes [3]

length-selectable switches (WSS) is being often used. The simplest WSS is a channel blocker, with a single input and output fibre, having the capability to power equalize or completely attenuate the WDM channels. The more capable 1×K WSS has a single input and K output fibres, adding the capability to independently route the individual WDM channels among the K fibres (Figure 16.). WSS with higher K requires a large micromirror tilting angle (>8°) and devices using vertical comb drive or double hinged angle amplification [10]. Gentler angle-bias response at large angles can be obtained by using alternative design that uses fringe electrical fields.

There are several other MEMS devices for optical networking applications such as polarization-mode dispersion (PMD) compensators, tunable laser, etc [11]. New developments in optical MEMS are based on materials technology and cost-effective processing. Optical MEMS are also benefiting from developments of IC industry such as BSOI technology that provided realization of low stress micromirrors, as well as production of other MEMS devices with reproducible mechanical properties and excellent planarity. Continuing progress results in products with better performances such as large-scale switches, variable attenuators, tunable

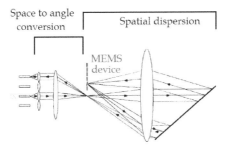

Figure 16. Schematic of MEMS 1×K WSS

filters, etc. High-voltage drivers and sense electronics are being integrated with highly reliable low-loss optical MEMS devices. Accuracy improvement of IC lithography and reactive ion etching provides necessary precision for optical MEMS production. Since there is a great scope for invention in MEMS device structure, materials and processing, optical MEMS will continue to play an increasingly important role in future of optical networks and ultra-high bandwidth communications.

It should be mentioned that besides functional optical MEMS devices, MEMS technology is also being applied in optoelectronic packaging. Ability to provide accurate passive alignment at low cost is one of the important assets of MEMS technology. MEMS approach provides accurate, low-loss optical connections between different guided wave optical components. Highly reliable connections realized using well characterized materials allow construction of complex interconnections. As an illustration, schematic of optical fibre fixed in a V-shaped groove by the triangular microclip is shown in Figure 17.

Figure 17. Schematic of optical fibre fixed in a V-groove by the triangular silicon nitride mechanical microclip

4. Reliability of optical MEMS

Reliability of optical MEMS for telecommunications is identified as the next manufacturers challenge for the forthcoming years due to a growing market and stricter requirements. Because of the vast diversity of device designs, materials and functions it is necessary to understand both technologies related variables as well as external variables such as environmental and operational conditions. MEMS reliability analysis is extremely important to identify and understand the different failure mechanisms that can be electrical or/and mechanical. Optical MEMS failure mechanisms are more complicated than those in micro-electronics for several reasons:

- MEMS devices are designed to interact with environment at various environmental conditions (e.g., temperatures),

- they are often hermetically sealed and they are expected to have long-term performances,

- some of the failures is impossible to predict,

- reliability testing for MEMS devices is not standardized unlike IC and microelectronics,

- for every new device new testing procedures need to be developed.

Design for test is important as well as performing parametric testing, testing during assembly, burn-in and final testing, testing during use, etc. Testing during assembly is of utmost importance for optical MEMS devices. It has two purposes. The first is to determine which devices are ready for the packaging process and the other is monitoring the yield of the packaging process. After the assembly devices are subjected to "burn-in" tests because packaged device may fail to perform due to the invasion of unwanted foreign substances such as dust particles and moisture. The main purpose of this test is to induce "infant mortality" failure on the manufacturing premises but not during operational lifetime (Figure 18). During the useful lifetime of the device the failure rate is relatively low. Failures are usually caused by external events such as vibration, shock, ESD, etc. Testing during use ensures proper functioning of the device for the intended application. Finally, device deteriorates due to intrinsic problems caused by material fatigue, frictional wear, and creep.

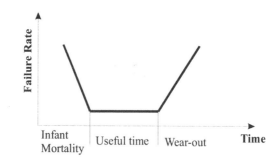

Figure 18. Failure rate as a function of time

One of the potential failure mechanisms of optical MEMS is stiction. Stiction occurs when surface adhesion at the contacting interface exceeds the restoring force. Adhesion may be driven by either capillary condensation or van der Waals forces [12]. Capillary condensation is affected by moisture and surface contamination, while van der Waals forces are affected by surface roughness. Since device dimensions are minute, gravity and other body forces do not play a significant role. Van der Waals forces are short range forces which cause materials to be attracted at the molecular level. The vulnerability to stiction can be significantly reduced by surface passivation coatings, the use of critical point (CO_2) drying of MEMS devices and moisture free packaging [10]. Enclosure in a controlled atmosphere and robust hermetic packaging greatly reduce the presence of moisture. Also, anti-stick layers are commonly being used to lower the surface interaction energy and prevent stiction. These layers provide hydrophobic surfaces on which water cannot condense and capillary stiction will not occur. However, the reliability and reproducibility of these layers is an important issue because of the high temperatures required in MEMS packaging process steps. The best way to avoid stiction failures is to eliminate presence of contacting surfaces by using adequate design or to enhance restoring force. In case of MEMS micromirrors, excessive adhesive force between the landing tip and its lending site may lead to stiction failure of the device. When the electronic reset sequence is applied, sufficiently high adhesive force may obstruct the movement of the mirror. Capillary water condensation causes the landing tip of the mirror and adequate landing site to become stuck. A partial vacuum is produced at the interface due to the surface tension and great forces are required to pull the tip and the landing site apart. For this reason, the usually used method for MEMS micromirror stiction elimination are implementation of springs on the landing tips of the mirror (Figure 19.) [13]. When the mirror landing tip lands on its landing site the spring bends and stores energy that will assist the mirror in taking off the surface when the reset pulse is being applied and bias voltage is being removed.

Figure 19. Stiction elimination: schematic of the spring tip and its landing site

Friction is another mechanism that impacts the lifetime of MEMS device. It is of interest when sliding/rotating optical MEMS are in question and it sets the upper limit of MEMS device lifetime. Friction occurs when two contacting surfaces move against each other. Repeated formation and breaking of contact lead to increase of the contacting stress. When the stress exceeds the material yield strength, material loss occurs. Significant wear finally causes mechanical failure. Frictional wear can be reduced by application of certain coatings (e.g. tungsten). Also, humidity can reduce wear by forming surface hydroxide but it can lead to increased stiction. However, elimination of rubbing surfaces during optical MEMS design phase is the best way to avoid friction [14]. Figure 20. shows an example of friction when optical MEMS devices are in question. A microengine in combination with the microtransmission is often used to drive a pop-up micromirror up, out of the plane. Sets of microgears provide linear motion with high degree of force. Intimately contacting surfaces repeatedly move against each other causing the augmentation of asperities that may lead to accumulation of debris and, finally, mechanical failure.

Figure 20. Schematic of the microengine affected by friction and wear

Lifetime of optical MEMS devices can also be affected by fatigue. Repeated motions can cause stress that even significantly below the crack strength, leads to crack growth and eventual failure. Crack growth can be facilitated by stress corrosion and for that reason is highly sensitive to humidity. Both silicon and polysilicon are not immune to fatigue. Stress engineering during design phase and materials selection can reduce the problem, but humidity control is the key factor to fatigue elimination. Micromirrors are often affected by fatigue. Each micromirror is hinged so it can rotate. Having in mind that each mirror will be switched thousands of times per second, hinge fatigue should be taken into consideration. In order to avoid fatigue, micromirror hinges are usually realized using thin-film technology. The fatigue properties of thin-film layers are different from those of bulk materials. Metal thin films exhibit much less fatigue than do their macroscopic counterparts since they do not have internal crystal structure because they are just a few grains thick [14]. Thin-films have less stiffness and therefore are less prone to breaking. Fatigue causes movement of dislocations to the surface of the material forming fatigue crack after enough damage has been accumulated. For that reason, not enough damage will accumulate on the thin film surface to form fatigue cracks. However, having in mind that the fatigue properties of thin films are often not known and that fatigue predictions are error prone, hinge structural materials should have material strength that far exceeds the maximum stress expected.

When strain varies with time under the constant stress, creep occurs. Movement of dislocations and diffusion of atoms trigger the deformation. It depends on the material in question, grain size, temperature and initial stress. Over the time surface flatness becomes affected by creep as well as parameters of mechanical parts. Metals are known to creep under stress, while silicon and polysilicon are more robust against creep as brittle materials. For optical MEMS devices silicon is often coated with a thin metal film. Reflective metal coatings on micromirrors are required for the desired optical performance. However, micromirors can become deformed during annealing. After cooling, micromirrors can have significant curvature due to the CTE mismatch between silicon and metal. When single sided metallization is in question, the curvature will slowly decrease as the metal creeps and not the underlying silicon. When symmetrical mirrors are in question, where both sides are metalized using two metal films deposited under different conditions, an uncontrolled increase in mirror curvature can be expected. By increasing silicon thickness flatter micromirrors can be obtained. However, that would affect the resonant frequency, response time and susceptibility to mechanical shock. It can also lead to very high drive voltages, with associated dielectric breakdown and dielectric charging issues [15]. When micromirror hinges are in question, unlike hinge fatigue, creep induced hinge memory poses a significant threat to MEMS micromirror device reliability. It is very significant life limiting failure mode that occurs when a micromirror operates in the same direction for a long period of time. When the bias voltage is removed the mirrors should return to a flat state. Their return to a non-flat state is known as a hinge memory effect (Figure 21.). The angle between the flat and non-flat state is called residual torque angle. As this angle increases, at one point the mirror will not be able to land to the other side. Main contributors to hinge memory failure are duty cycle and operating temperature, but the main cause of this type of failure is the creep [12]. As structural mirror beam materials high melting point compounds are being used such as Al_3Ti, AlTi, AlN because high melting point metal often has low creep. Since it is obvious that temperature affects the lifetime of the micromirror device, thermal management is very important. In order to keep temperature in the device within the acceptable range, heatsinks are being used. Adequate thermal management significantly influences lifetime of the device allowing the mirrors to be efficiently controlled over a longer period of time.

Figure 21. Schematic presentation of the hinge memory failure mode

Common cause of electrical failure when MEMS devices are in question is anodic oxidation on unpassivated silicon wiring and electrodes. Positively biased electrode oxidizes under the high humidity. Negatively biased electrode remains unaffected. In order to eliminate anodic oxidation the primary goal is moisture elimination by using hermetically sealed packages. Also, for any silicon used as conductors, passivation should be provided.

Environmental robustness is a great reliability concern for all MEMS devices. Examination of micromirror environmental robustness is based on standard semiconductor test requirements such as temperature cycling, thermal shock, moisture resistance, ESD, cold and hot storage life, etc. Similar to ICs, MEMS devices are also susceptible to ESD damage. Sudden transfer of charge that occurs between MEMS device and person or piece of equipment causes ESD damage when on-chip protection circuits are not available because of the incompatibility to IC processing or design complexity (Figure 22.). ESD proof clothing and tools are obligatory when elimination of ESD in MEMS fabrication is in question.

Figure 22. Schematic of ESD damage: polysilicon comb finger

Another large optical MEMS reliability concern is vibration [16, 17]. Due to the sensitivity and fragile nature of many MEMS, external vibrations can have disastrous implications. They may cause failure through inducing surface adhesion or through fracturing device support structures. Long-term vibration can also contribute to fatigue. Another issue can be shock. Shock is a single mechanical impact instead of a rhythmic event. A shock creates a direct transfer of mechanical energy across the device. Shocks can lead to both adhesion and fracture. Although optical MEMS devices seem fragile due to their small size, their size proved to be one of their greatest assets. Small size enables their robustness. They proved to be able to sustain low-frequency vibrations and mechanical shock without damage. However, besides being an asset, size may be related to another type of failure mechanism. Dimensions of MEMS devices are so small that the presence of the smallest particle during fabrication may cause non-functionality of one or more devices (Figure 23.). For that reason the source of each contaminating particle should be detected and eliminated, especially during packaging, because particles sealed in the package may affect operation of the device during its lifetime. Hermetic packaging can provide adequate protection, electronic contacts and, if necessary, interaction with the environment through the window transparent to light. Also, vacuum packaged devices eliminate effects of capillary stiction. Failure due to contaminations introduced during packaging is the most common failure mode of optical MEMS devices.

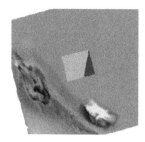

Figure 23. Schematic presentation of micromirror failure caused by particle contamination

5. Conclusion

Optical MEMS devices are still relatively unproven in telecommunications applications and the most optical MEMS devices are not yet fully qualified. A brief insight in reliability of optical MEMS devices for telecommunications applications has been presented in this chapter. Several major reliability issues have been disused: stiction, friction, fatigue, creep, etc. However, developing reliable optical MEMS component is non-trivial. Production of reliable optical MEMS device requires sophisticated design considerations and better control of microfabrication processes that are used in realization of MEMS device. One of the challenges is providing temperature insensitive, particle free, mechanically stable environment. Usually submicron alignment tolerances are required and high port count optical MEMS require handling and packaging of large numbers of optical fibers, micromirrors, lenses and electrical control leads. Light collimation and focusing, wavelength separation, precisely controlled, large, flat and highly reflective microstructures, significant control electronics are just some of the issues. Reliable packaging is an imperative. Reliable package must not prevent mechanical action of moving parts of the structure, but it should prevent transfer of heat, moisture, outgassing, etc [18, 19]. Another issue is the need for credible testing techniques applicable during fabrication, assembly and packaging, as well as during operational lifetime of the device. As the number of ports grow testing requirements become challenging since multiple, expensive laser sources and flexible test architectures are required. Besides all that, competing technologies pose significant threat to optical MEMS applications (Table 3.) Micromotors, LCD devices, planar waveguides, solid state technologies such as Lithium Niobate and Semiconductor Optical Amplifiers (SOA) can realize various wavelength and fiber management component functions although many coincide that 3D optical MEMS is the only all optical technology that can integrate such complex switching functions in a small package. The key to successful future of optical MEMS in telecommunications market lies in improvement of device structure, materials and processing. Lower losses are required that can be obtained through flatter micromirrors and better quality lenses. More ports are required to handle the expansion of the traffic and reliable and cost-effective packaging is needed to house thousands of tiny fragile MEMS structures. It should be pointed out that industrial standardization of MEMS technology is at least several years away [17, 20] and till then optical MEMS devices will be custom made according to customer requirements. The lack of information flow, as well as reluctance in sharing experience will keep optical MEMS devices from full commercialization although there are several commercially successful applications.

Technology	Cost	Perf.	Scale	Reliab.	Integ.	Maturity
MEMS	strong	good	strong	not determined	strong	moderate
Micromotors	weak	strong	weak	moderate	weak	strong
LCD	weak	strong	moderate	weak	weak	good
Planar Waveguide	good	good	moderate	good	strong	moderate
Solid state /SOA	weak	moderate	moderate	not determined	good	weak

Table 3. Comparison of the Technology Alternatives for Wavelength Management/Fibre Management components [21]

Acknowledgements

Authors are grateful for the partial support of the Ministry of Education, Science and Technological Development of Republic of Serbia (contracts III45007 and III44003).

Author details

Ivanka Stanimirović and Zdravko Stanimirović

IRITEL A.D., Belgrade, Republic of Serbia

References

[1] Fedder GK. Mems Fabrication. In Proceedings of International Test Conference, ITC 2003, ISBN: 0-7803-8106-8, 30 September – 2 October 2003, 1: 691 – 698, IEEE 2003.

[2] Moore DF, Syms RRA. Recent developments in micromachined silicon. Electronics & Communication Engineering Journal, ISSN: 0954-0695, December 1999, 11(6) 261 – 270, 1999.

[3] Syms RRA, Moore DF. Optical MEMS for telecoms. Materials Today, ISSN: 1369-7021, July/August 2002, 26-35, Elsevier Science Ltd 2002.

[4] Judy JW. Microelectromechanical systems (MEMS): fabrication, design and applications. Smart Materials and Structures, ISSN: 0964-1726, December 2001, 10(6) 1115-1134, IOP Publishing Ltd 2001.

[5] Wu MC. Micromachining for optical and optoelectronic systems. Proceedings of the IEEE, ISSN: 0018-9219, November 1997, 85(11) 1833 – 1856.

[6] El-Fatatry A. Optical Microsystems, Mechno-Optical-Electro-Mechanical Systems – MOEMS. RTO-EN-AVT-105 - MEMS Aerospace Applications, ISBN 92-837-1113-0, February 2004, NATO Science and Technology Organization, 2004.

[7] Noell W, Clerc PA, Dellmann L, Guldimann B, Herzig HP, Manzardo O, Marxer C, Weible K, Dandliker R, de Rooij N. Applications of SOI-based optical MEMS. IEEE Journal of Selected Topics in Quantum Electronics, ISSN: 1077-260X, January/February 2002, 8(1) 148 – 154.

[8] De Dobbelaere P, Falta K, Gloeckner S, Patra S. Digital MEMS for optical switching. IEEE Communications Magazine, ISSN: 0163-6804, March 2002, 40(3) 88-95.

[9] Lin LY, Goldstein EL. Micro-electro-mechanical systems (MEMS) for WDM optical-crossconnect networks. In Military Communications Conference Proceedings, MIL-

COM 1999, ISBN: 0-7803-5538-5, 31 October-03 November 1999, Atlantic City, NJ, USA, 2: 954 – 957, 1999 IEEE.

[10] Chang CP. Optical MEMS Technology for Telecommunication. In Proceedings of 7th International Conference Solid-State and Integrated Circuits Technology, ISBN: 0-7803-8511-X, October 18-21, 2004, 3: 1981 – 1986, 2004, IEEE.

[11] Choo H, Muller RS. Devices, Structures, and Processes for Optical MEMS. Transactions on Electrical and Electronic Engineering, 2007, 2:216-231, IEEJ 2007

[12] Tadigadapa S, Najafi N. Reliability of Microelectromechanical Systems (MEMS). In Proceedings of Reliability, Testing, and Characterisation of MEMS/MOEMS Conference, ISBN: 0-8194-4286-0, October 22-24 2001, 197 - 205, San Francisco, CA, USA, SPIE, Bellingham, USA 2001.

[13] Douglass MR. DMD Reliability: a MEMS success story. Reliability, Testing, and Characterization of MEMS/MOEMS II, Ramesham R, Tanner DM, (eds), Proceedings of SPIE, Vol. 4980(2003), SPIE 2003.

[14] Van Spengen WM. MEMS reliability from a failure mechanisms perspective. Microelectronics Reliability, ISSN: 0026-2714, July 2003, 43(7) 1049–1060.

[15] Hartzell AL, da Silva MG, Shea HR. MEMS Reliability.New York, London, Springer 2011.

[16] Hsu TR. Reliability in MEMS packaging, In Proceedings of 44th International Reliability Physics Symposium, ISBN: 0-7803-9498-4, San Jose, CA, March 26-30, 2006, IEEE International.

[17] Hsu TR. Introduction to Reliability in MEMS Packaging. In Proceedings. of International Symposium for Testing and Failure Analysis, San Jose, California, November 5, 2007.

[18] Stanimirović I, Stanimirović Z. Reliability of MEMS. In: Takahata K. (ed) Micro Electronic and Mechanical Systems, Rijeka:InTech, 2009, p.177-184.

[19] Najafi K. Micropackaging Technologies for Integrated Microsystems: Applications to MEMS and MOEMS. Micromachining and Microfabrication Process Technology VIII, Yasaitis JA, Perez-Maher MA, Karam JM. (eds), Proceedings of SPIE, Vol. 4979 (2003), SPIE 2003.

[20] Tadigadapa S, Najafi N. Developments in Microelectromechanical Systems (MEMS): A Manufacturing Perspective. Journal of manufacturing science and engineering, ISSN: 1087-1357, 2003, 125(4) 816-823, American Society of Mechanical Engineers, New York, USA, 2003.

[21] Robinson SD. MEMS Technology – Micromachines Enabling the "All Optical Network". In Proceedings of 51st Electronic Components and Technology Conference, ISBN: 0-7803-7038-4, 29 May - 01 Jun 2001, Orlando, Florida, USA, 2001 IEEE

Chapter 5

CMOS Compatible Bulk Micromachining

John Ojur Dennis, Farooq Ahmad and M. Haris Khir

Additional information is available at the end of the chapter

1. Introduction

The last two decades have seen the emergence and prevalence of Micro-Electro-Mechanical Systems (MEMS) technology. The reduction in size, low power consumption and low cost is the ultimate goals of application specific MEMS devices. One way to achieve these goals is the monolithic integration of MEMS technology with the standard integrated circuit (IC) technology. The utilization of the conventional IC technology as a platform for design of the fast-growing MEMS technology has led to development of numerous devices and technologies in the past years. Much effort has been made and large capital has been invested into mainstream Complementary Metal-Oxide Semiconductor (CMOS) compatible MEMS technology. This approach has enabled MEMS devices to be directly integrated with CMOS circuits, allowing smaller device sizes and higher performances. This integration of MEMS technology with the mainstream CMOS technology is referred to as CMOS-MEMS technology [1].

1.1. Bulk micromachining technologies

Among all micromachining technologies, Bulk micromachining is the oldest one. To make small mechanical components using bulk micromachining technology, substrate material such as silicon wafer is selectively etched. There are two major classes of bulk micromachining: wet bulk micromachining (WBM) and dry bulk micromachining (DBM).WBM gives nice feature resolution and vertical sidewalls for deep trenches in the substrate but have numerous disadvantages such as incompatibility with microelectronic circuits and higher cost due to mask requirement. The fabrication cost, time and simplification of micromachining for MEMS devices is reduced by using dedicated MEMS foundry and making the prototyping develop-ment cycle much shorter by reducing number of masks and photolithography steps. CMOS MEMS technology is a good solution for post micromachining because in this technology CMOS circuitry is protected completely by the top metal layer and the same top metal layer is used as mask to define MEMS structures in MEMS region during the post-CMOS micromachining. That is why this post micromachining process is known as CMOS-compatible.

1.1.1. Wet bulk micromachining

Wet bulk micromachining is done normally with the help of chemicals. In chemical wet etching, the silicon substrate is immersed into the solution of reactive chemicals which etches the exposed regions of substrate at measureable rates. Wet etching typically uses alkaline liquid solvents, such as potassium hydroxide (KOH) or tetramethylammonium hydroxide (TMAH) to dissolve silicon which has been left exposed by the photolithography masking step. These alkali solvents dissolve the silicon in a highly anisotropic way, with some crystallographic orientations dissolving up to 1000 times faster than others. Fig. 1 shows the three major crystal planes in a cubic unit of silicon.

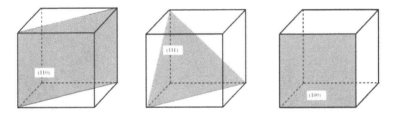

Figure 1. The three orientations <110>, <111> and <100> are the respective directions normal to the plane

For example in silicon anisotropic etching, planes <111> are etched at slower rate than all other planes. The reasons for the slow etch rate of planes <111> are, high density of silicon atoms exposed to the etchant solution in this direction and three silicon bonds laying below the plane. Fig. 2 (a) shows a schematic and (b) three dimensional photo of a typical wet anisotropic etching of a silicon substrate.

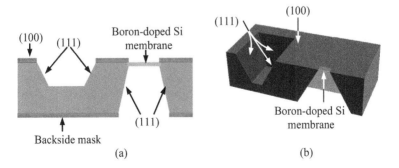

Figure 2. a) 2-D Schematic and (b) 3-D diagram of a typical wet bulk micromachining of Si

Wet chemical micromachining technology is popular in MEMS industry because of high etch rate and selectivity. There is one drawback and that is the mask is also etched during normal etching process so it is suggested to find a mask that doesn't dissolve or at least dissolves at a much slower rate than the silicon substrate. In wet etching, etch rates and selectivity can be modified by various methods such as: (a) chemical composition of etching solution, (b) dopant concentration in the substrate, (c) adjusting temperature of etching solution and finally (d) Modifying crystallographic planes of the substrate.

Chemical wet etching in bulk micromachining may be further subdivided into two parts: isotropic wet etching where the etch rate is not dependent on crystallographic orientation of the substrate and etching proceeds in all directions at equal rates as shown in Fig.3 (a) and anisotropic wet etching where the etch rate is dependent on crystallographic orientation of substrate as shown in Fig. 3 (b). Often in order to control etching process and uniform etch depths across the wafer, etch stops are used. Generally three types of etch stops are used in micromachining: Dopant etch stops, Electrochemical etch stops and Dielectric etch stop.

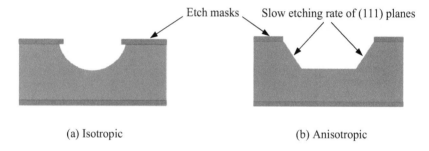

Etch masks   Slow etching rate of (111) planes

(a) Isotropic   (b) Anisotropic

Figure 3. Difference between (a) isotropic and (b) anisotropic wet etching

Usually in order to control etching process and uniform etch depths across the wafer, etch stops are used. Generally three types of etch stops are used in micromachining: Dopant etch stops, Electrochemical etch stops and Dielectric etch stop [2].

1.1.2. Dry bulk micromachining

Dry etching in bulk micromachining is classified into three groups: Reactive ion etching (RIE), Vapor phase etching and Sputter etching. Reactive ion etching (RIE) uses both physical and chemical mechanisms to achieve high levels of resolution. The process is one of the most diverse and most widely used processes in industry and research. Since the process combines both physical and chemical interactions, the process is much faster. The high energy collision from the ionization helps to dissociate the etchant molecules into more reactive species. In the RIE-process, cations are produced from reactive gases which are accelerated with high energy to the substrate and chemically react with the silicon. The typical RIE gasses for Si are CF_4, SF_6 and $BCl_2 + Cl_2$.

Both physical and chemical reaction is taking place with the physical part similar to the sputtering deposition. If the ions have high enough energy, they can knock atoms out of the material to be etched without a chemical reaction. It is very complex task to develop dry etch processes that balance chemical and physical etching, since there are many parameters to adjust. By changing the balance it is possible to influence the anisotropy of the etching with the chemical part being isotropic and the physical part highly anisotropic and therefore the combination can form side walls that have shapes from rounded to vertical.

A special subclass of RIE which continues to grow rapidly in popularity is deep RIE (DRIE). In this process, etch depths of hundreds of microns can be achieved with almost vertical sidewalls. The primary technology is based on the so-called "Bosch process", named after the German company Robert Bosch which filed the original patent, where two different gas compositions are alternated in the reactor. The first gas composition etches the substrate, and the second gas composition creates a polymer on the surface of the substrate. The polymer is immediately sputtered away by the physical part of the etching, but only on the horizontal surfaces and not the sidewalls. Since the polymer only dissolves very slowly in the chemical part of the etching, it builds up on the sidewalls and protects them from etching and working as pasivation, as illustrated in fig. 4. In fig.4 (a) SF_6 is etching the silicon while in fig. 4 (b) C_4F_8 is playing the role of passivation and in fig. 4 (c) again SF_6 is etching. As a result, etching aspect ratios of 1 to 50 can be achieved. The process can easily be used to etch completely through a silicon substrate, and etch rates are 3-4 times higher than wet etching. In the following part more details are given on DRIE. Under our MEMS research group in Universiti Teknologi PETRONAS, high aspect ratio of 3 to 50 μm and complete etch through is achieved as shown in fig.5.

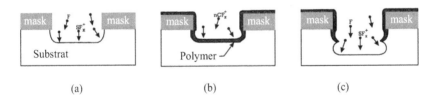

| (a) | (b) | (c) |

Figure 4. Illustration of etching and passivation in DRIE

Vapor phase etching (also called Chemical dry etching) does not use liquid chemicals or etchants. This process involves a chemical reaction between etchant gases to attack the silicon surface. The chemical dry etching process is usually isotropic and exhibits high selectively. Anisotropic dry etching has the ability to etch with finer resolution and higher aspect ratio than isotropic etching. Due to the directional nature of dry etching, undercutting can be avoided. Some of the ions that are used in chemical dry etching is tetrafluoromethane (CH_4), sulfur hexafluoride (SF_6), nitrogen trifluoride (NF_3), chlorine gas (Cl_2), or fluorine (F_2) [3]. Vapor phase etching can be done with simpler equipment than what RIE requires. The two most common vapor phase etching technologies are silicon dioxide etching using hydrogen fluoride (HF) and silicon etching using xenon diflouride (XeF_2), both of which are isotropic in nature.

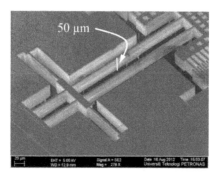

Figure 5. FESEM image of a high aspect ratio plasma etching using DRIE with inhibitor

Usually, care must be taken in the design of a vapor phase process to not have bi-products formed in the chemical reaction that condense on the surface and interfere with the etching process.

Sputter etching is essentially RIE without reactive ions. The systems used are very similar in principle to sputtering deposition systems. The big difference is that the substrate is now subjected to ion bombardment instead of the material target used in sputter deposition [4].

2. CMOS compatible micromachining

CMOS compatible micromachining is now commonly known as CMOS MEMS technology. During post processing in CMOS MEMS technology, the top metal layer acts as mask to define MEMS structures. For the front side processing of the die no photolithographic steps are required. CMOS MEMS technology facilitates the wiring of MEMS structures to integrated circuits due to many metal layers. Through monolithic integration of the CMOS-MEMS devices with circuits, the parasitic capacitances is reduced and overall high performance devices is achieved. Dry post CMOS MEMS technologies are divided into two classes: Thin film micromachining technology (Thin-film Post-CMOS MEMS Technology) and DRIE bulk micromachining technology (DRIE Post-CMOS MEMS Technology).

2.1. Thin-film post-CMOS MEMS technology

The process flow of thin-film CMOS-MEMS with three metal layers is shown in Fig. 6. The CMOS circuit region is designed to be covered by the top metal layer and the MEMS structures are also pre-defined by the top metal layer or the other interconnect metal layers. Fig.6 (a) shows the cross-section of the original chip after CMOS foundry fabrication, with a passivation layer on top. Two processing steps are performed only on the front side of the chip. First, the pre-defined MEMS structure is opened by etching the SiO_2 dielectric stack layer as shown in Fig.6 (b). This is done by an anisotropic SiO_2 plasma etch using Tegal SS110 A DRIE system based on Bosch technology [5]. Next, an anisotropic silicon DRIE is performed using the same Tegal DRIE system as shown in Fig.6 (c), followed by isotropic silicon etch, which releases the MEMS structure by undercutting the silicon beneath as shown in Fig.6 (d). The depth of the anisotropic etch into the silicon controls the gap between the released structure and the silicon substrate. This gap should be large enough to eliminate the parasitic capacitance between the MEMS structures and the silicon substrate. In practice, it is normally on the order of 30 μm [6].

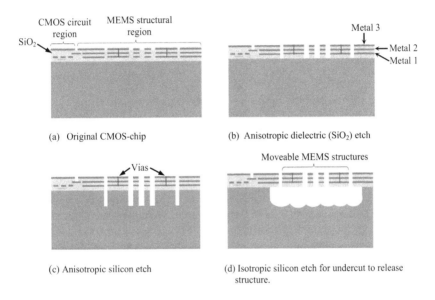

(a) Original CMOS-chip

(b) Anisotropic dielectric (SiO$_2$) etch

(c) Anisotropic silicon etch

(d) Isotropic silicon etch for undercut to release structure.

Figure 6. The process-flow for thin film DRIE post-CMOS micromachining

A resonant magnetic field sensor and piezoresistive accelerometer are examples of devices fabricated using this technology [7, 8]. This simple fabrication process yields much smaller parasitic effects as compared to the other processes such as poly MUMPs and it provides flexible wiring by using the multiple metal layers. However, there are some drawbacks in this technology, which limit the performance of the fabricated accelerometers. First, there is large vertical curling in the comb fingers of the suspended MEMS structures caused by the residual stress gradient existing in the composite SiO$_2$/Al layers as shown in Fig. 7. Additionally, due to different thermal expansion coefficients (TEC) of Al and SiO$_2$, the curled MEMS devices exhibit a strong temperature dependence, which limits their utility. Although particular compensation technology was employed to reduce the structure mismatch by using a specially designed frame, the device fabricated using this thin film process still has a stringent size limit.

Second, for sensors which require proof mass like inertial sensors fabricated using this technology, the requirement of holes on the proof mass reduces the proof mass, resulting in a lower mechanical sensitivity of the sensors. To achieve a capacitance change large enough for

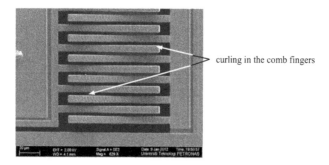

curling in the comb fingers

Figure 7. FESEM image of typical comb fingers fabricated using thin film CMOS-MEMS technology

the readout circuit to detect, the dimension of the sensors may need to be considerably large, which is in conflict with the dimensional limit imposed by the possibility of structure curling. Third, in-plane curling of the thin film structures due to fabrication variations also limits the maximum size of the device.

Although there are some specific processes designed for low residual stress thin films [9], Farooq et al utilized the Malaysian Institute of Microelectronic Systems (MIMOS Bhd) 0.35 μm CMOS foundry processes to reduce the curling of the MEMS structures by proper design of the structures as described in [10]. Fig. 8 shows the successfully released central shuttle of magnetic field sensor fabricated using this process. It can be clearly observed that the curling in the structure has been significantly reduced as indicated by the minor curling in the fingers. In addition to the collection of technological data directly from the employed CMOS foundries, systematic post-CMOS process calibration must be conducted to characterize the process variations and their effects on the MEMS devices.

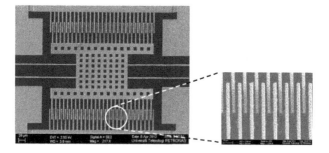

Figure 8. FESEM image showing comb fingers of the central part of a magnetic field sensor

Lastly, since the last release step is an isotropic etch, it undercuts the silicon close to the circuit and structure anchors. The ratio of the vertical and lateral etching rate is approximately 3:5. To protect the silicon underneath the circuit region from being etched away during the structure release, the CMOS circuitry must be placed far away from the microstructures, especially when a large separation between the microstructures and substrate is needed. As a result, significant chip area is wasted due to the protection margin around the MEMS structures. Since the silicon underneath the mechanical anchors of the MEMS structures is also etched away, the suspension of the mechanical structures is softened, which results in a lower mechanical performance and less robustness of the device [10].

2.2. DRIE post-CMOS MEMS technology

To reduce the disadvantages of the above thin-film post-CMOS MEMS process, DRIE post-CMOS MEMS technology is developed to incorporate bulk single crystal silicon (SCS) into the MEMS structures that are formed by CMOS thin films. Due to the improvement in DRIE technology, high aspect ratio CMOS MEMS structures have been exhibited as shown in Fig. 9. The achievement of maximum aspect-ratio in DRIE system is an important factor in the MEMS structure design [5]. It decides the dimensional limit of the structures fabricated using that DRIE system. Once a lateral feature to be etched is fixed, the maximum thickness is uniquely defined.

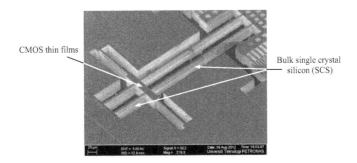

Figure 9. FESEM image showing the incorporation of bulk SCS to the CMOS thin films.

2.2.1. Process flow

The DRIE post-CMOS MEMS technology branches from the thin film technology described in the previous section. DRIE CMOS-MEMS provides an approach to implement thick and flat MEMS structures with improved mechanical performance and device strength due to advantage of both flexible wiring with multiple metal layers in CMOS technology and the capability of high aspect ratio etching. Fig.10 shows the process flow of the DRIE post-CMOS MEMS technology.

To define the MEMS structure thickness (thin film plus substrate), the process starts with backside silicon DRIE etching by the selective application of photoresist at the back-side as shown in Fig.10 (a). The maximum thickness of this structure is limited by the smallest etching pattern on the front side of the MEMS structure and the maximum etch aspect-ratio. Next, to

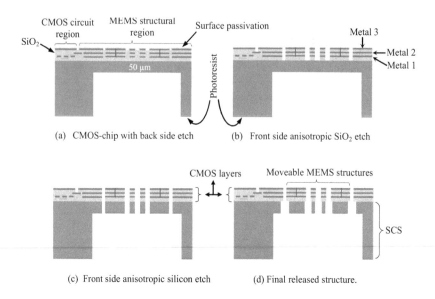

(a) CMOS-chip with back side etch   (b) Front side anisotropic SiO₂ etch

(c) Front side anisotropic silicon etch   (d) Final released structure.

Figure 10. The process-flow for DRIE CMOS micromachining

define the MEMS structures, an anisotropic SiO_2 etch is performed on the front side of the chip as shown in Fig. 10 (b) similar to the thin film process. In the following step, the structure is released by etching through the remaining SCS substrate. This step differs from the thin film process in that an anisotropic DRIE, instead of isotropic Si etch, is performed to release the structure as shown in Fig.10 (c). Flat and large MEMS structures can be obtained by incorporating SCS underneath the CMOS interconnect layers because the residual stresses in the SiO_2/Al thin-films are mitigated by the thick SCS, showing very little out-of-plane curling as shown in Fig. 11 where it can be seen that the sense fingers are perfectly flat and aligned.

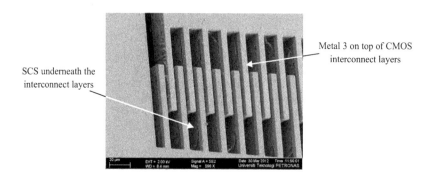

Figure 11. FESEM image showing comb fingers with thick SCS underneath

In some MEMS structures isotropic silicon etch can be added to create nice mechanical structures which only consist of CMOS thin films. A resonant magnetic field sensor fabricated under University Technology PETRONAS (UTP) MEMS Research group incorporates both CMOS thin-film layers for the beams and silicon substrate along with CMOS thin films in the side torsional coils as shown in Fig. 12.

Thin film CMOS
layers as beams

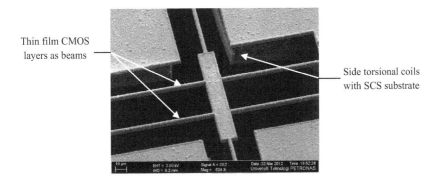

Side torsional coils
with SCS substrate

Figure 12. FESEM image of CMOS thin layers and incorporation of SCS with thin layers as MEMS structures

Double side alignment is required in the backside DRIE step to define the thickness of the MEMS structures. It is observed that normally the remaining silicon for bulk MEMS structures is on the order of tens of microns and by the substrate thinning there is no apparent circuit performance degradation. Therefore, there is no hard and fast constraints for alignment in the backside etch. A reliable sensing and actuation is achieved with the flat MEMS microstructures enabled by the DRIE CMOS MEMS technology. A resonant magnetic field sensor with piezoresistive sensing is fabricated using this technology [11]. The technology is very suitable for the fabrication of, in particular, thermally actuated micromirrors where bimorphs are used to elevate the mirror plate. Numerous resonant chemical sensors, electrothermal micromirrors and accelerometers have been fabricated using this technology [12].

2.2.2. Design rules for MEMS devices

For CMOS circuit designs, there are different ways to break up a layout into basic elements. Designers usually divide and analyze a circuit by the function of individual components, devices and interconnect. The devices may be active (e.g. transistors, diodes) or passive devices (e.g. resistors, capacitors and inductors). The interconnects include in-plane wires, pads and vias which connect multiple metal layers. The critical information on the geometric patterns includes the size of each feature, the spacing between features and the overlap of features. These geometric data, required in the design, leads to evolutions of processes and ensures successful fabrication of microelectronic circuitry. In order to fulfill the requirements of successful MEMS device designs, processing information, which quantitatively describes the capabilities and limitations of the process under predetermined processing conditions and parameter settings, is necessary and crucial. This information is abstracted as MEMS design rules and is verified in the design environment as illustrated in Fig. 13.

Figure 13. Design rules function between MEMS designers and MEMS fabrication engineers.

This section explains how to construct concise and conclusive design rules. We used MIMOS Berhad Malaysia standard 0.35 μm CMOS design rules for the fabrication of CMOS MEMS devices and only the design rules for mechanical structures are discussed in this section. These MEMS design rules add further constraints to layouts. Similar to CMOS design rules, a complex micromechanical layout is fragmented into small pieces, and MEMS design rules contain only fundamental geometries directly related to the post CMOS micromachining process. A set of geometric constraints and design rules are then based on the final process.

Design rules are expressed in the design-rule file used in layout tools such as Cadence Virtuoso, Magicor, Mentor Graphic, layout tool etc. After all processing completion, a post-CMOS micromachined chip can be partitioned into a released area (containing free standing parts and free moving parts) and a non-released area (anchored parts) according to the mechanical behavior. The objective of post CMOS processing is the final result of releasing microstructures and maintaining critical dimensions. There are practical difficulties in monitoring each intermediate process stages. Therefore characterization of the process to extract design rules is performed at the end of the entire process flow. An overview of a typical post CMOS MEMS magnetic sensor is shown in Fig. 14.

The anchored part is the area outside of the red dash-line box on FESEM, whereas the released part is the area inside the red border. The close-up FESEM view of circled portion is shown in Fig. 14 (b). Sufficient paths for the gas flow are required in order to allow the reactive radicals in the plasma to undercut MEMS structures. These paths can surround mechanical structures, so that the undercut can happen from the periphery of the structures. The gas flow can also pass through structures via etch release holes. Fig. 14 illustrates various situations for reactive species undercut: from one side of the structures toward the anchor region, from both sides of beam structures, or outward through etch release holes on plates.

Theoretically, any shape of the feature which opens a gas flow path from the top of the chip to the substrate is suitable for the purpose of undercut. However, currently in CMOS foundries, Manhattan layout is preferred, or at least convenient. The majority of CMOS foundries only accept patterns with edges which are parallel, perpendicular or 45° with respect to the horizontal/vertical alignment lines. Non-Manhattan features require more and smaller rectangular patterns to approximate the geometry, and therefore, require a larger data base and longer exposure time, which leads to higher cost. Today, raster-scanning mask-making

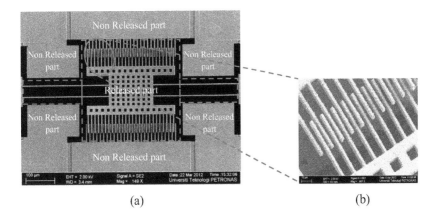

(a)   (b)

Figure 14. Partitioning of a MEMS sensor for design-rule checking

technologies such as e-beam or laser systems can overcome the difficulties mentioned above but non-Manhattan shapes approximated by a smaller raster-beam size still require more time in fabrication [13]. Following the practice of CMOS fabrication, this work only explores rectangular design rules. The major purpose of this section is to ensure the release of moving/ free-standing parts and to guarantee the integration of anchor parts. Other process related rules are not included. According to situations described in Fig. 14, there have been three features encountered in the MEMS structure layout for the purpose of undercut: large openings, gaps and holes. To control the etch rates in these opening, dummy structures are used as shown in Fig.15 (a) while figure 15 (b) shows the fully released structure.

Test structures are used to extract design rules. A large opening undercut rule specifies the maximum undercut that can be achieved by the process. Any microstructure wider than this value will not be released. This rule also specifies the minimum size requirement of the anchor layout, and the skirt for bonding pads. Bonding pads are usually on the periphery of the chip for easy access during bonding and no crossing of bonding wires on the top of the chip. Therefore, at least one side of bonding pads faces large opening and is lateral undercut during the release. The sites that experience forces during wire bonding should avoid these undercut regions and bonding pads need extra area to compensate this lateral undercut. Moreover, this large opening rules specifies the minimum space required from the edge of the etch pit to preserve Si area containing circuitry during micromachining. One way to obtain this data is to use FESEM to measure the undercut from the periphery of the chip. Another way is to use an optical microscope to check which size of square pattern has been released as illustrated in Fig. 16 (a).

The undercut value is half the size of square width. A gap undercut rule is required for releasing MEMS structures such as a comb finger or a spring. The test structure for this rule is a set of cantilevers with the undercut mainly determined from the sides of the beams. To follow the situation in real layout, such as comb fingers, multiple cantilever beams with equal space

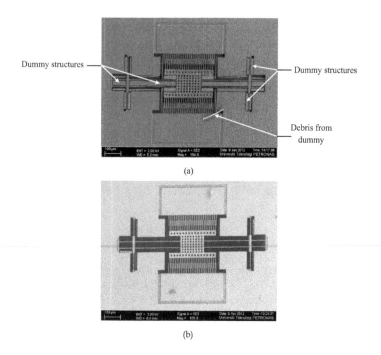

(a)

(b)

Figure 15. FESEM images of (a) unreleased device showing dummy structure at a large etch holes and (b) fully released structure.

between them are used as shown in Fig. 16 (b). The beam width (w) represents the width of the structure to be released. The gap (g) represents the separation between the structures. The beam should be long enough that the undercut from the tip of the cantilever doesn't dominate the etch. Since the residual stress induces out-of-plane curling upon release, the release can be easily detected by an interferometric measurement. A hole undercut rule is required to release plate structures. The test structure for this design rule is a set of plates with release holes. These plates are anchored by four thin beams as shown in Fig. 16 (c). Such plates should be large enough to ensure that the release of the plate is determined by the etch release holes and not from the periphery of the plate.

Similar to the cantilever test structures, the curling after the release of plates can be detected by the interferometric measurement. Extra processing time requires layouts to be modified such that the circuitry is far away from the edge of the mask, all the anchors are wide enough, and all bond pads have enough margins for bonding and probing. The spacing between the squares and the spacing from a square to the surrounding anchor mask area are the size of the square. This is to ensure that squares are uniformly etched from all directions. For each group of cantilever test structures, as in Fig. 16 (b), the widths of all cantilever beams are the same value. The spacing between beams and from tips of cantilever beams to anchor mask areas are the same value [14].

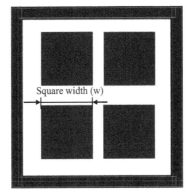

(a) The square structure illustrating the large opening undercut rule

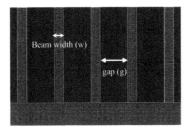

(b) The cantilever structure illustrating the gap undercut rule

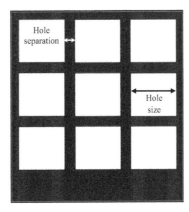

(c) The plate structure illustrating the hole undercut rule

Figure 16. Layout of three test structures used for extracting MEMS design rules for post-CMOS micromachining.

3. DRIE based CMOS-MEMS devices

In this section, two devices fabricated using CMOS-MEMS bulk micromachining technology are described. The devices are a resonant MEMS magnetic field sensor and a resonant MEMS chemical sensor.

3.1. Resonant MEMS magnetic field sensor

The post-CMOS process steps of the resonant magnetic field sensor using DRIE were designed to successfully release the sensor structure. The first step in the post-CMOS process was at wafer level with the selective application of photoresist at the back-side of the 8" inch silicon wafer to etch from the back-side to produce a substrate thickness of approximately 50 μm as shown in Fig.17. DRIE was performed using SS110A Tegal plasma etcher to anisotropically etch the silicon substrate to the desired thickness.

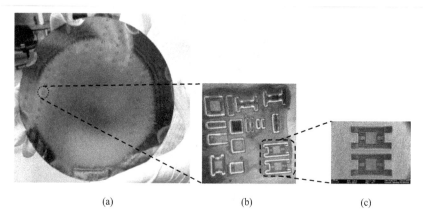

| (a) | (b) | (c) |

Figure 17. a) Optical image of back side Si etching at wafer level, (b) Backside optical image of a chip and (c) FESEM of backside of a sensor.

Optical microscope was used to estimate the thickness of the Si substrate during the etching process. Fig.17 (c) shows the back-side view of one of the sensors that has been successfully etched to approximately 284 μm depth leaving a about 50 μm thickness under the CMOS layers. From the back-side etching of the bulk silicon, the silicon etching rate was found to be ~3.66 μm/min. The second step was dicing the wafer as shown in Fig. 18 (a). Dicing of the wafer has to be done before the front side RIE of SiO₂ and DRIE of silicon is implemented. This sequence was followed in order to prevent the breakage of the released structures.

The third process was SiO₂ RIE on the diced chip shown in Fig. 18 (b), which was performed from the front-side of the chip using Tegal plasma etcher. Front side RIE process opened the pattern of the resonant magnetic sensor by removing the SiO₂ layer and thus exposing the silicon underneath while the MEMS sensor pattern was maintained by metal 3 acting as a mask

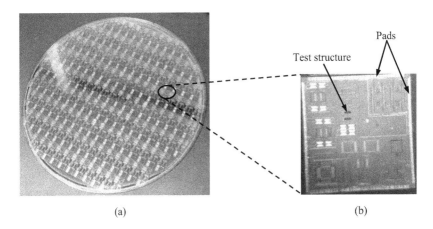

(a)   (b)

Figure 18. a) Optical image of the diced wafer and (b) optical image of front side of a 5 mm × 5 mm chip.

for the pattern. The process at chip level started by flipping over the chips such that the thin film was now on the front side. The chips were then attached to the 8" platinum coated carrier wafer using kapton tape as illustrated in Fig.19. After 13 minutes of etching, a gray colour (silicon colour) appeared in the trenches that indicated the completion of SiO_2 etching. The test structures to the right of the sensor beam were used to estimate SiO_2 etching rate and to avoid over-etch of the SiO_2 material as shown in Fig.18 (b). The total SiO_2 etching depth was approximately 5 μm.

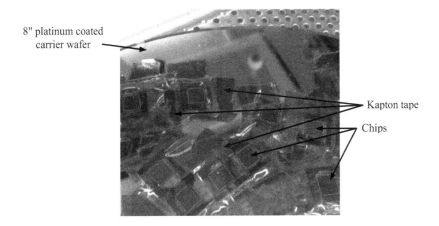

Figure 19. Sample preparation for front side etching

Next, front-side bulk silicon DRIE process step was performed using the etching rate of 3.66 μm/min obtained from previous backside DRIE etching of silicon. The thickness of the central shuttle and long beams was approximately 50 μm, which was observed under optical microscope during back-side etching. Therefore the test structures on the chip were used as reference. Fig.20 shows FESEM micrograph of the fabricated sensor with inset showing a close-up of the comb fingers and part of dummy structure close to the beams that has fallen out while others still remain intact on the substrate [10]. Additional 3 more minutes were required to fully etch-through. This happened when all the dummy structures dropped and indicated that the etched-through process was completed.

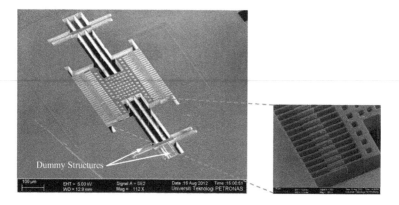

Figure 20. FESEM image of the fabricated CMOS-MEMS resonant magnetic field sensor with inset showing shuttle and stator fingers.

3.2. Resonant MEMS chemical sensor

In resonant MEMS chemical sensors, known as gravimetric sensors, the principle of detection of the gaseous species is based on the change in resonant frequency of the microresonator membrane or plate. This frequency change results from a change in the mass of the microresonator due to absorption/adsorption of an analyte molecule onto the surface of the active material deposited on it. Fig. 21 shows 2-D schematic diagram of the chemical microsensor device.

The post-CMOS micromachining process steps of the resonant MEMS chemical sensor were similar to the process steps of the resonant MEMS magnetic field sensor. The process started with the selective application of photoresist at the back-side of the die around the sensor followed by sample placement on the platinum coated carrier wafer. The second step of the process was to perform back-side silicon DRIE to achieve proof mass thickness of approximately 40 μm. The same plasma silicon etcher was used to anisotropically etch the silicon substrate to the desired thickness. Fig. 22 shows the FESEM image of the successfully released MEMS chemical sensor with the inset showing a close-up view of the perfectly flat sensing comb fingers with the SCS underneath [15].

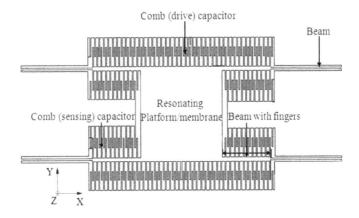

Figure 21. D Schematic of the resonant MEMS chemical sensor

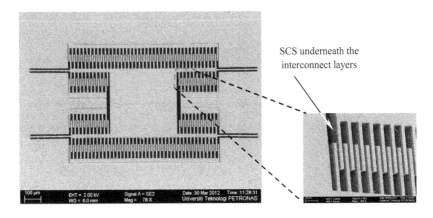

Figure 22. FESEM image of fabricated device with inset showing a close-up view of the fingers with the SCS underneath

4. Conclusion

This chapter discussed bulk micromachining technology with particular emphases on DRIE post CMOS MEMS bulk micromachining. The chapter was divided into three sections. In the first section an introduction to bulk micromachining of silicon and isotropic and anisotropic wet and dry etching was given. The second section discussed briefly DRIE post-CMOS micromachining process with particular emphasis on DRIE post-CMOS bulk micromachining process and the third and last section provides a few examples of devices fabricated by our research group using the DRIE CMOS-MEMS process. These devices were resonant MEMS magnetic field sensor and resonant MEMS chemical sensor. The aim of the chapter was to discuss and analyze practical processes involved in the design of micromechanical devices using 0.35 μm CMOS technology.

Acknowledgements

The authors would like to thank MIMOS Bhd Malaysia for facilitating the microfabrication of the sensors and MOSTI Malaysia for financially supporting this research under E-Science project No.04-02-02-SF0095.

Author details

John Ojur Dennis, Farooq Ahmad and M. Haris Khir

Department of Fundamental and Applied Sciences, Department of Electrical and Electronic Engineering, Universiti Teknologi PETRONAS, Perak, Malaysia

References

[1] Hongwei Qu and Huikai XieProcess Development for CMOS-MEMS Sensors with Robust Isolated Bulk Silicon Microstructures", *IEEE/ASME Journal of Micro-Electro-Mechanical Systems*, , 16, 1152-1161.

[2] http://www.memsnet.org/about/fabrication.html

[3] http://www.meengr.ccny.cuny.edu/bridge/material/joined%20files.pdf

[4] http://www.mems.ac.cn/jishuwenxian/(2009). Etching_Processes.html

[5] Xie, H, Erdmann, L, Zhu, X, Gabriel, K. J, & Fedder, G. K. Post-CMOSprocessing for high-aspect-ratio integrated silicon microstructures," Journal of Microelectromechanical Systems, (2002). , 11, 93-101.

[6] Fedder, G. K, Santhanam, S, Reed, M. L, Eagle, S. C, Guillou, D. F, Lu, M. S. C, & Carley, L. R. Laminated high-aspect-ratio microstructures in a conventional CMOS process," The Ninth Annual International Workshop on Micro Electro Mechanical Systems, San Diego, CA, (1996). , 13-18.

[7] Ahmad, F, Dennis, J. O, Khir, M. H. M, & Hamid, N. H. A CMOS MEMS Resonant Magnetic field Sensor with differential Electrostatic actuation and Capacitive sensing," Advanced Materials Research Journal, vols 403-408, (2012). , 4205-4209.

[8] Haris, M, Qu, H, & Cmos-mems, A. Piezoresistive Accelerometer with Large Proof Mass", IEEE NEMS (2010). Xiamen, China, Jan 2010.

[9] Xie, H. Gyroscope and micromirror design using vertical-axis CMOS-MEMSsensing and actuation," Ph. D. Thesis, Department of Electrical and Computer Engineering, Carnegie Mellon University, Pittsburg, PA, (2002).

[10] Ahmad, F, Dennis, J. O, Khir, M. H. M, & Hamid, N. H. Lorentz force based CMOS Resonant Piezoresistive Magnetic Sensor with differential electrostatic actuation" MY. Patent, Jan.16, (2012).

[11] Farooq AhmadDesign and simulation of mechanical behavior of MEMS-based resonant magnetic field sensor with piezoresistive output", (2010). International Conference on Mechanical and Electrical Technology, 09/2010

[12] Fedder, G. K. Tri-axial high-g CMOS-MEMS capacitive accelerometer array", (2008). IEEE 21st International Conference on Micro Electro Mechanical Systems, 01/2008.

[13] Zhu, X. Post-CMOS Micromachining of Surface and Bulk Structures," Ph. D. Thesis, Department of Electrical and Computer Engineering, Carnegie Mellon University, Pittsburg, PA, (2002).

[14] Brand, O. CMOS-Based Resonant Sensors", IEEE Sensors (2005).

[15] Ahmed, A. Y, Dennis, J. O, Khir, M. H. M, Saad, M. N. M, & Method, A. Apparatus and Sensor for Reversible Detection of Analyte Molecule" MY. Patent, Dec.16, (2011).

Chapter 6

Mechanics of Nanoelectromechanical Systems: Bridging the Gap Between Experiment and Theory

Hamed Sadeghian, Fred van Keulen and
Hans Goosen

Additional information is available at the end of the chapter

1. Introduction

Nowadays, mechanical designers and engineers of elastic structures at ultra-small scales face an interesting challenge; traditional flexures with several components, mechanical joints/ welds and linkages are almost impossible to manufacture with existing microsystem fabrication technologies. Therefore, the majority of mechanical components, elements and building blocks are based on micro/nano machined elastic flexures [1]. Despite the complexity of modeling and analyzing these systems, their design uses the averaged relation between stress and strain, which requires a relatively accurate knowledge of their effective elastic properties, specifically the effective elastic modulus. Experimental results show that the elastic properties are constant at length scales of meters down to micrometers [2]. However, in order to increase the performance, i.e. sensitivity and dynamic range, the dimensions of mechanical devices have been scaled down towards a few nanometers. Consequently, high performances such as single-electron tunneling [3], sub-attonewton force sensing [4], and sub-femtometer displacement sensing [5] have been successfully achieved. Unlike at micron and higher length scales, at sub-micron and nanometer length scales the effective elastic behavior shows strong scale-dependent behavior, meaning that the elastic properties are no longer constant, but a function of length scale. For all the application examples above, the high performance was achieved due to the mechanical response of a nanosystem, which strongly depends on the effective elastic properties. Therefore, a clear understanding of the scale-dependent behavior is important for the design and performance of nanosystems.

1.1. Scale-dependence

The existence of scale-dependent behavior has been confirmed by experimental measurements, including resonance frequency tests [6], tensile testing in scanning electron microscope

(SEM) [7], transmission electron microscope (TEM) [8, 9], atomic force microscope (AFM) [10, 11] and nanoindenter [12] and also theoretical investigations, including *ab initio* and density functional theory (DFT) [13, 14, 15], molecular dynamics (MD) [16, 17, 18, 19] and modifications to continuum theory [20, 21, 22, 23]. Although scale-dependence has been observed by both theory and experiment, a considerable discrepancy still remains between the experiments and models. As an example, in mono-crystalline [110] silicon nanowires and nanocantilevers the experimental scale-dependence has been observed at about 150 nm [24, 11, 8, 6, 25], whereas theoretical studies do not agree with any scale-dependence larger than 10 nm. In order to highlight the inescapable discrepancy between experiment and theory, the results of recent experimentally measured effective Young's modulus $\hat{E}$ and those extracted from theories for silicon nanocantilevers and nanowires were collected from relevant literature and are illustrated in Fig. 1. The figure demonstrates that a large gap exists between the theoretical predictions and experimental observations. The smallest silicon nanowire with a diameter of about 12 nm that has been experimentally tested by Han *et al.* [26] shows an effective elastic modulus which is about one third of the bulk value, while the largest theoretically modeled (mainly atomistic simulations) silicon nanowire (diameter ~ 10 nm) shows an effective elastic modulus that is closer to the bulk value [26]. It can also be observed from Fig. 1, that there is lack of data, both in experiment and theory; 1) there is no computational data for structures in the range from 10 to 150 nm, so no direct comparison between experiments and theory at about 100 nm can be performed. The reason is that it is computationally extremely expensive and impossible to model the atomistic systems as large as the ones experimentally tested. Although quasi-continuum approaches have been developed as computationally efficient methods [27, 28], they are lacking important phenomena, such as surface reconstruction, defects emitted from surfaces and surface stress induced phase transformations. Thus, one of the biggest challenges in the field of nanomechanics is the development of a multi-scale modeling framework, capable of performing simulations at various length scales. 2) There is no experimentally measured data below 12 nm, therefore for structures smaller than 10 nm a direct comparison between theory and experiment cannot be performed either. The main reason is that it is extremely challenging and difficult to perform nanoscale experiments with sufficient accuracy and resolution, while introducing minimum disturbances to the measured nanostructure [24]. Therefore, experimental capabilities need to be improved, enabling measurements of structures with a size of less than 10 nm.

Moreover, the observed discrepancies are not only between the simulations and experiments but also between various experimental measurements. The observed discrepancies [25, 29, 24] could be explained by: some exterior factors such as ill-defined boundary conditions and uncertainties in experiments [30], loading modes (extensional vs. bending) [31, 23], calibration and required input accuracy (i.e. variations in geometry) limitations [11] as well as intrinsic nature of inaccuracy in the experimental methods (i.e. mass effects in resonance based tests) [32, 33, 34] and the environmental effects such as surface contaminations, measurement induced-errors and native oxide layers could all be influencing the measurement at sub-micron scales, inducing a large disagreement between measurement data. On the other hand, there are intrinsic effects which become significant at sub-micron scales such as surface stress, surface elasticity, nonlinear bulk effects, surface reconstructions, crystal defects and fabrication induced imperfections that make the scale-dependent phenomena more complicated.

Due to their small sizes, nanosystems predominantly have high surface-to-volume-ratio, thus, unlike micron size systems, the influence of surface stress cannot be neglected. Therefore, the scale-dependence of elastic behavior is generally attributed to surface effects, including surface elasticity [35, 20, 23] and surface stress [36, 37], surface oxidation [31, 29, 38] and surface contamination [39]. Besides the intrinsic surface effects, it was speculated that surface geometry and cross section are important factors. Most simulations have been performed on nanowires with rectangular cross section, which is significantly simplified compared to actual cross sections that are closer to rhombic or pentagonal configurations [40]. In order to examine to what extent these effects are significant, McDowell *et al.* [31] using atomistic simulations, studied the scale-dependent elastic behavior of nanowires with various cross sections and with various surface steps. The conclusion was that, although these factors influence $\hat{E}$ to some extent, they alone are insufficient to explain the experimentally observed trends.

In addition to surface geometry, the loading method can also influence $\hat{E}$. The extracted $\hat{E}$ in the extensional mode is different from the one extracted in the bending mode [16]. Consequently, there are two types of experimental characterizations: those measuring $\hat{E}$ in the extensional mode, such as uniaxial tensile loading tests [41, 42, 9], and those in the bending mode such as resonance frequency based [6] tests or bending tests inside an atomic force microscopes [10, 11]. Most of the research today focuses on the bending mode of $\hat{E}$ because it is more sensitive to surface stress and surface elasticity effects and because of its importance in sensing and actuating applications. Again the question arises: at which scale does the difference between bending and extensional modes become significant? To answer this question, McDowell *et al.* [16] investigated the effect theoretically and found out that it is influential only for dimensions below approximately 8 nm. This indicates that other factors beyond the above effects play a role in experimentally observed nanosystem scale-dependent elastic behavior. Therefore, the question remains: why does $\hat{E}$ start to decrease already at sizes between 10 nm and 150 nm?

A couple of issues have been raised by researchers in order to reduce the gap between the experiments and the theory [24, 11, 39, 16, 25].

A recently recognized substantial effect, which has not yet been considered, is accounting for the fact that experimentally measured nanosystems are not perfect (such as existence of surface defects, native oxide layers, contaminations etc.) and not defect free (such as single crystal defects, fabrication induced imperfections etc.) in contrast to the perfect mono-crystalline structures studied using atomistic simulations [19, 17, 14, 13]. The goal of this chapter is to try to determine to what extent these effects influence the effective elastic modulus of nanosystems, to try to reduce the existing gap between the experiments and theory of scale-dependence, and to determine the origin of the scale-dependence in silicon nanosystems.

1.2. Outline of the chapter

This chapter is organized as follows: Section 2 describes the fabrication process for micro/nano cantilevers, used for experimental investigations of scale-dependent $\hat{E}$, including the inspection of the fabricated structures. In Section 3 the experimental determination of $\hat{E}$ is presented and discussed. Section 4 is focused on the theoretical investigations of surface effects and native

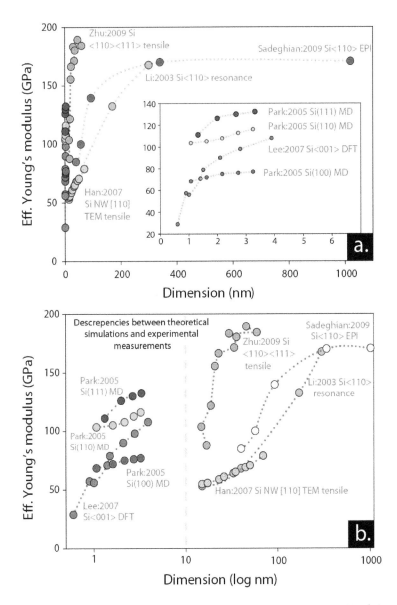

Figure 1. (a) Illustration of the scale-dependence of mono-crystalline silicon effective elastic modulus $\hat{E}$ obtained via computations and experimental observations [29]. The inset shows results for the scale less than 10 nm, indicating that all simulations are limited to scales of less than 10 nm due to complexity and expense of computations. (b) A logarithmic plot of (a) to further illustrate the differences between the simulation and the experimental results. Both show sharp drops near the down-scaling side; the question is why discrepancies exist? (Reprinted with permission from [29]. Copyright 2010 IOP Publishing Ltd.)

oxide layers on the scale-dependent $\hat{E}$, including molecular dynamics calculations as well as semi-continuum approaches, to show to what extent these effects are responsible for the observed scale-dependence. Section 5 discusses the possible effects of defects that may contribute to the reduction of $\hat{E}$ at larger scales. In order to show the extent that defects play a role in scale dependence, molecular dynamics is utilized to quantify the effects of defects on $\hat{E}$. Finally in Section 6, a summary, conclusions and recommendations for future work on scale-dependence of $\hat{E}$ are presented.

2. Fabrication process and inspection of mono-crystalline silicon micro/ nanosystems

Micro and nanocantilevers, double clamped structures and plates are widely fabricated with top-down approaches [6, 32], while nanowires and nanotubes are produced invoking bottom-up techniques [43]. This section describes the fabrication process and inspection of the silicon cantilevers and double clamped beams studied in this chapter. The characteristic of the acquired structures are 1) their thickness has to be accurately known, 2) their surface has to be very smooth and 3) the pre-bending due to releasing the structures, or in other words, the gap between the substrate and the structures has to be known. Inspections of the test structures were conducted in order to evaluate these criteria. The thickness of the device and substrate layers were measured by ellipsometry. AFM roughness measurement together with the results of ellipsometry were used for investigating the roughness of the structures. And finally, white light interferometry was used to measure the initial curvature of the structures.

The mono-crystalline silicon cantilevers and double clamped beams were fabricated on (100) Silicon On Onsulator (SOI) wafers from Soitech, using the Smartcut® process to form the buried oxide and the silicon device layer. The used wafers are shipped with 1 µm (measured 1019 nm) thick buried oxide and a 340 nm thick low boron-doped silicon device layer. From these original wafers, thicker layers were obtained by epitaxial growth, while for the thinner ones they were thermally oxidized and then etched to the desired thicknesses. The surface roughness measured by AFM revealed a smooth surface (RMS of ~ 0.095 nm) on both the original wafer, as well as the epitaxied sample and the oxidation-etched samples (Fig. 2).

The device thicknesses were controlled by careful timing of the epitaxy and the oxidation processes. Test wafers were added to the processing batch to verify the thicknesses; the modified SOI layers were then precisely determined by ellipsometry, a pre-defined model is fitted onto the measured data to obtain the exact thicknesses. Typical fittings and their results are shown in Fig. 3. Samples with desired thickness were first spin coated with HMDS as a primer that serves as an adhesion promoter for the photoresist. The recipe is: 5000 RPM, 2 min at 200°C bake in a hot plate, then coat with photoresist AZ5214 of 1.25 µm, 5000 RPM, 2 min at 120° C and then pattern in a lithography step (Fig. 4.a).

The wafers were developed in pure MF321 for about 85 seconds and 30 seconds in water. The sample is then etched in a SF6 based plasma etcher to pattern the top silicon layer until reaching the BOX layer (Fig. 4.b). To release the cantilever, the residue of photoresist is removed in nitric

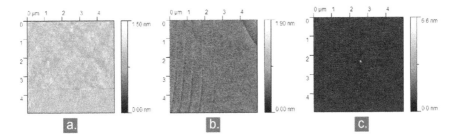

Figure 2. AFM roughness measurement of the surfaces of (a) the epitaxied 1 µm, (b) the original 340 nm and (c) the etched 40 nm top silicon device layer. The smooth surface indicates continuous single crystal growth/etch from the original surface.

acid and the BOX is etched in Hydrofluoric (HF) solution, in order to release the structures (Fig. 4.c). Drying the structures directly after etching with HF and diluting in water can cause stiction of the structures to the substrate due to the water surface tension. In order to prevent device stiction, freeze drying or critical point drying (CPD) were used. The cantilevers were 170 to 8 µm long, 20 to 8 µm wide and 1019 to 40 nm thick. Fig. 5 shows SEM images of fabricated cantilevers and double clamped beams.

Due to the isotropic nature of the release-etching, undercuts and hidden anchors were formed at the clamping site of the cantilevers. This has an overall lengthening effect on the device. The extra length was determined with the optical microscope, as shown in Fig. 6.a. In addition, as shown in the HF etching of the BOX, the bonding surfaces of the SOI wafers were in the middle of the BOX layer about 340 nm below the top silicon oxide interface (Fig. 6.b and Fig. 6.c). This eliminates the bonding-induced defects on the top silicon and greatly improves the surface quality of the device silicon. Residual stress, due to the inconsistency of both the thermal expansion coefficient and the crystal lattice period between the substrate and the thin film, is unavoidable in surface micromachining techniques [44]. The compressive residual stress in relatively thin double clamped beams can create pre-bending. Moreover, it can create built-in moments, which in released cantilevers cause them to curl out of plane. The cantilever has one end free and therefore, it can partially release the stress. The consequence of this stress releasing is an upward or downward curvature of the cantilever.

Bending profiles of each structure before the experiment were measured using a Wyko NT3300 optical surface profiler that utilizes white light interferometry for high resolution three-dimensional (3D) surface measurements. Fig. 7.a shows the white light interferometry measurements for a cantilever which is approximately flat. One of the most important parameters that affects the accuracy of experiments is the curvature of the structures. Therefore, for each cantilever that is bent during the releasing process, a profile measurement was performed. This bending modifies the initial gap between the suspended structure and the substrate. Fig. 7.b and Fig. 7.c show the reconstructed 3D profiles of a cantilever and a double clamped beam with significant curvatures. Fig. 7.d shows the profile measurements in the length direction of the curled cantilever. The markers indicate the length of the cantilever.

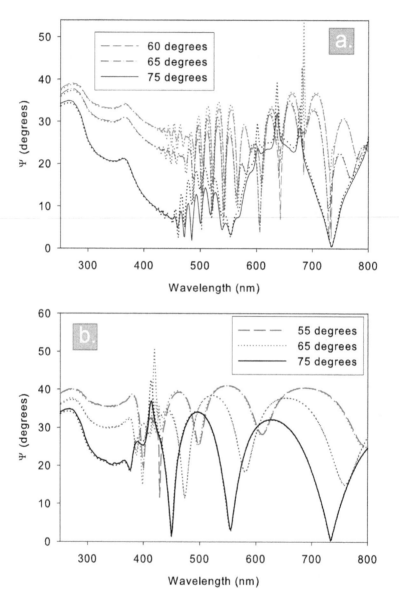

Figure 3. Typical ellipsometer measurements of the SOI wafers with (a) 40 nm and (b) 1019 nm device layer thicknesses. The dots are the measured data, solid lines represent the expected values generated from the fitted thickness with different incident angles (angels are shown in the legend). The mean squared errors (MSE) of the fitting were less than 18.

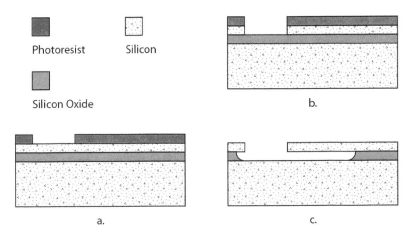

Figure 4. Fabrication process of the cantilevers. (a) photoresist is patterned. (b) Using the photoresist as an etching mask, the silicon underneath is etched with SF6 plasma. (c) Residue of the resist is stripped in nitric acid and the device is submerged in HF solution to etch the oxide. The device is then put in a critical point drier (CPD) to release the suspending cantilever structure.

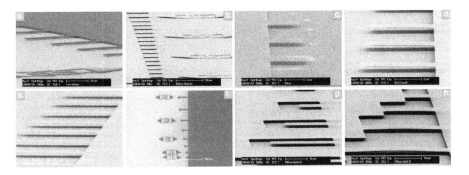

Figure 5. SEM of (a) 1019 nm, (b) 340 nm, (c) 93 nm, (d) 57 nm, (e) 40 nm thick cantilevers. SEM of (f) paddle-shaped cantilevers, (g) covered cantilevers and (h) double clamped beams, respectively.

3. Experimental measurement

Meaningful experimental measurements of scale-dependent elastic behavior from micro-down to nano-scales have been shown to be one of the major challenges in the field of experimental micro/nanomechanics [24, 45]. As a consequence, a majority of research concerning the scale-dependent elastic response has focused on developing experimental methods to study the phenomena [24, 41, 42, 8, 32]. One of the most common approaches to measure $\hat{E}$ is based on the resonance response of the system. $\hat{E}$ is extracted via a simple Euler-Bernoulli beam equation from a resonance frequency response acquisition [6, 46]. The main advantages of this method, which make it popular and widely used, are 1) ease of use; besides measuring

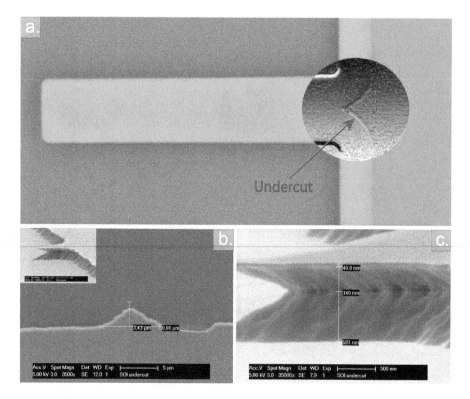

Figure 6. (a) Optical microscope image of a cantilever with hidden anchor due to under etching of the base of the cantilever. (b) SEM of a hidden anchor. The cantilever was forcibly removed in order to see the hidden anchor. The inset shows the bonding surfaces of the SOI wafers were in the middle of the BOX layer. (c) Close-up view of bonding interface and silicon device layer.

the resonance frequency no additional modifications on the system are required, and 2) the method is very fast, while being applicable in various environments, including vacuum, gaseous and liquids. The method has some limitations, especially when being used at nano-scale. The experimental results include errors due to the uncertainty of the boundary conditions [30]. The non-ideal boundary conditions lead to a lower resonance frequency, which leads to lower estimates of $\hat{E}$. Moreover, the resonance frequency depends on both the stiffness and the mass [39, 33]. It is therefore difficult to decouple solely by a resonance response the stiffness from the mass changes caused by surface contamination, native oxide and other adsorbed layers [33, 39, 34]. The smaller the nanosystem, the bigger this effect will be. In order to show the extent to which the effects of the extra mass due to contamination or native oxide are significant, we investigated the effects of surface contaminations on the resonant behavior of the cantilevers fabricated in Section 2. We have shown that the mass and the stiffening effects of the contaminations can cause significant shifts in the resonance frequency of the cantilevers, and, thus, the subsequent estimate of the effective stiffness.

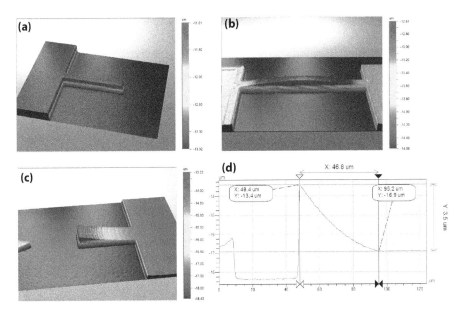

Figure 7. Reconstructed 3D image of (a) a cantilever which is approximately flat, (b) a curved double clamped beam and (c) a curled cantilever using white light interferometry. (d) curvature profile measurement of the cantilever in (c).

The resonance frequency of the cantilevers was measured with an optical laser detection setup equipped with a vacuum chamber [39]. The cantilevers are in resonance due to Brownian motion caused by thermal noise. The resonance frequency and its shift were obtained from the Lorentzian fit of the frequency-bandwidth curves [39]. If the surface of the cantilever is clean without any contamination, such as adsorbed water and gas molecules, then the theoretical effective Young's modulus of the cantilever can be calculated as [39]

$$\hat{E} = \frac{0.0261 f_0^2 L^2 \rho}{t} \tag{1}$$

where f_0 is the resonance frequency, L is the length of the cantilever, ρ is the density of the cantilever material and t is the thickness, respectively. In order to determine the variations in $\hat{E}$ associated with the surface contamination, the resonance frequency of cantilevers with contaminations and with minimized contaminations were compared. Assuming that the contamination is distributed evenly along the cantilever, the variations in $\hat{E}$ can be calculated using (only for small variations)

$$\Delta\hat{E} = 2\frac{\hat{E}}{f_0}\Delta f \tag{2}$$

where $\Delta\hat{E}$ is the variation in $\hat{E}$, and Δf is the variation in the resonance frequency due to the contaminations. To show experimentally the effect of ambient contamination, we analyzed the resonance frequency shift of cantilevers by the following procedure (similar to the work in [39] for inaccuracies in frequency-based mass sensors). After fabrication of the cantilevers, their resonance frequencies, f_{bv}, were measured in ambient air, to allow for the adsorption of different gases and water molecules and their saturations. At this stage, the masses and stiffness from surface contaminations are added to the newly fabricated cantilevers and are referred to as the added mass-stiffness effects. The cantilevers were then placed into a vacuum (10^{-6} mbar) for a sufficiently long time to allow for degassing and desorption at the surfaces of the cantilever (reduced added mass-stiffness effects). Further, the resonance frequencies were measured in a vacuum (f_v). Then, the vacuum chamber was vented with ambient air and the resonance frequencies were measured immediately after the pressure reached atmospheric pressure (f_{av}). The measurements were done within a few minutes in order to ensure minimal re-adsorption, while providing the same amount of air mass loading compared to the measurements in the first step (f_{bv}). This procedure was repeated for 1019, 340 and 93 nm thick cantilevers. A typical resonance frequency measurement with the procedure above is illustrated in Fig. 8 for a 93 nm thick, 8 μm wide and 8 μm long cantilever. A 20.9 KHz resonance frequency shift was observed between the resonance frequency before and after vacuum. For thicker cantilevers the resonance frequency shift was less, due to their lower surface-to-volume ratio and consequently their being less sensitive. Table 1 shows the results of the resonance frequency measurements, their shifts and, consequently, the variations in $\hat{E}$ extracted from equation 2.

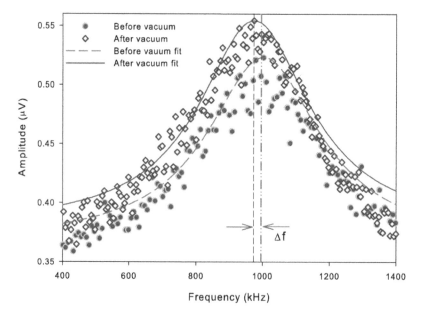

Figure 8. The influence of surface contamination on the resonance frequency of a 93 nm thick silicon cantilever.

The shift in frequencies is caused by the combination of both the additional adsorbed mass and the stiffening effect on the cantilever [39]. Therefore, the true stiffening effects of the contamination cannot be obtained by simple subtraction of the shifts in the resonance frequencies. The results from Table 1 clearly show that the experimental measurements of the scale-dependent $\hat{E}$ via the resonance frequency method lead to a considerable number of uncertainties.

Lxwxt (µm³)	f_{bv}	f_v	f_{av}	$\Delta f = f_{av} - f_{bv}$	$\Delta\hat{E}$ (GPa)	$\Delta\hat{E}/\hat{E}$ (%)
96 × 8 × 1.019	123.65	147.968	128.85	5.2	0.80	8.4
24 × 8 × 0.340	583.73	607.108	585.48	1.75	0.30	0.6
8 × 8 × 0.093	994.53	1090	973.6	-20.93	-3.90	-4.2

Table 1. The resonance frequency of cantilevers in air and vacuum and the variations in extracted $\hat{E}$. The resonance frequency data are in KHz.

Another popular method which is used to characterize the scale-dependence is the bending tests with the use of force spectroscopy in a scanning probe microscope (SPM) [11]. SPM has very high force and displacement resolution and sensitivity, yet it has significant uncertainties in its measurement and interpretations. Measuring the absolute deflection requires a painstaking calibration of the deflection sensor, as well as the risk of tip slippage [24] and indentation of the probe tip [41]. Another important factor which degrades the accuracy of the method is the lack of knowledge in the precise force. In a typical SPM measurements, the force can only be obtained from a calibrated cantilever with known spring constant. The spring constant of an SPM cantilever usually deviates from its designed nominal value by a significant amount. The lack of an accurate spring constant calibration method, together with the above issues result in inaccuracies which can amount to more than 26% error [11].

In order to avoid the aforementioned issues in experimental investigation of the scale-dependence, we implemented the recently developed measurement method, which uses the quasi-static electrostatic pull-in instability (EPI) phenomenon [32]. The distinctive advantage of the EPI lies in its well-known sharp instability and the possibility of applying an SI-traceable force along the length of the beam [32]. The details of the phenomenon can be found elsewhere [44]. Due to the quasi-static nature of the method, it does not suffer from mass-loading effects. The beauty of the methods lies also in its simplicity and ease of use. The required actuation voltage can be precisely measured using standard electrical test equipment and a microscope. Fig. 9 shows a schematic illustration of the EPI setup. It consists of a voltage source and a semiconductor testing probe station. The pull-in voltage was measured by slowly increasing the voltage difference between the cantilever and the substrate. Visual observation of the color changes and the sudden snap-in of the cantilever were used to determine the pull-in voltage.

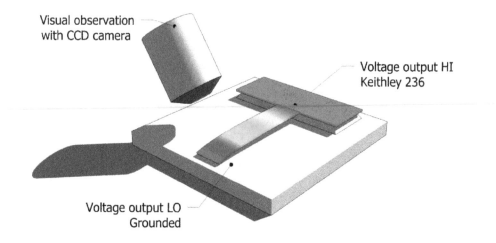

Figure 9. Schematic view of the measurement setup for EPI. The driving voltage on the cantilever is applied through a probe contact and the substrate is grounded. Reprinted with permission from [32]. Copyright 2010 American Institute of Physics.

For each cantilever, the pull-in voltage and geometry of various cantilevers with different lengths were measured. Fig. 10 shows the experimental measured pull-in voltages versus the length of cantilevers with different thicknesses. By solving the nonlinear differential equation of the electromechanical system, knowing the pull-in voltage and the geometry of the cantilever, one can extract $\hat{E}$ for each thickness [32]. Fig. 11 shows the extracted $\hat{E}$ as a function of the cantilever thickness. Error bars were calculated according to [32]. The maximum calculated error due to uncertainties in the cantilever's geometry measurements and the measured voltage was 12% [32]. The main error contributions comes from the geometry, and specifically, the length. Results in Fig. 11 show a considerable scale-dependent behavior in the effective stiffness for bending, $\hat{E}$.

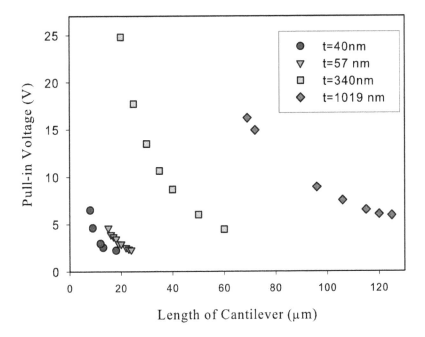

Figure 10. Measurement results of pull-in voltage versus the length for an array of cantilevers with different thicknesses and lengths.

Above 150 nm the effective modulus is constant and converges to the bulk value of [110] silicon elastic modulus. Somewhere near 150 nm it starts to decrease monotonically. It can be seen that at 40 nm, the $\hat{E}$ is about half of the bulk value. So the question to be answered is: what causes this scale-dependent behavior?

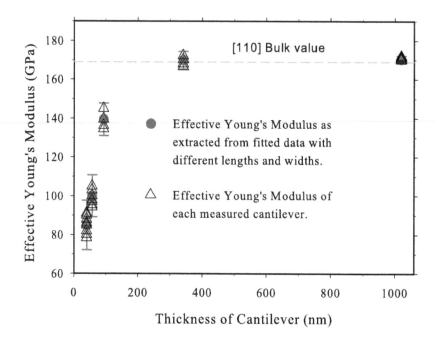

Figure 11. The scale-dependent *É* for bending. Reprinted with permission from [32]. Copyright 2010 American Institute of Physics.

4. Theoretical investigations: Impact of surface effects and native oxide layers

The influence of surface effects at scales higher than micrometers on the overall elastic behavior is negligible and the elastic behavior can be explained by the bulk properties. However, in nano systems, due to their small size, the surface-to-volume ratio is quite large and therefore, the surface effects cannot be neglected. This has motivated major research to try to explain the scale-dependence by the surface effects [41, 47, 20, 17]. The surface effects can be categorized into two types; the extrinsic effects such as surface contamination, native oxide layers and environmental adsorptions, and the intrinsic effects which originate from the difference in the atomic configurations near or at the surface and in the core of the bulk. The latter are known as surface stress and surface elasticity effects. The energy associated with the atoms near the surface is different from the energy of atoms in the core. The surface atoms have lower coordination numbers and electron densities and therefore tend to adopt equilibrium lattice spacing differently from the bulk ones [48, 23]. On the other hand, the epitaxial relationship from bulk to surface has to be maintained, therefore, bulk atoms strain the atoms near the surface and create the so-called surface stress [13, 23, 36, 49]. Consequently, the surface atoms like to reconstruct and can deviate easily from their original ideal situations, and therefore, have different elasticity compared to the bulk, known as surface elasticity [41, 47, 20, 23]. In order to predict the surface effects on $\hat{E}$, we have developed a semi-continuum, two-dimensional plane-stress framework [23], which takes into account the influence of surface elasticity on the elastic behavior of nanocantilevers. One of the main modifications to the previous models [47, 20] is the inclusion of coupling stiffness that arises from a surface stress and surface elasticity imbalance caused by either surface reconstruction [50] or molecular adsorption [51] at the top and bottom surfaces [22]. According to this framework, the bending mode of $\hat{E}$ can be calculated as

$$\hat{E} = E_b \left(1 + \frac{3\Sigma S}{E_b t} \right) \tag{3}$$

where E_b is the bulk value of Young's modulus, known to be 169 GPa for [110] mono-crystalline silicon, ΣS is the sum of the surface elasticity at the top and the bottom surfaces and t is the thickness of the cantilever. In order to calculate $\hat{E}$ the contribution of the surface elasticity (ΣS) is required. Because the surface region is only a few atomic layers thick, atomistic calculations are necessarily involved in order to calculate the contribution of the surface elasticity, and in general to include the surface effects in the modeling of nanosystems. We carried out molecular dynamics calculations (MDC) using a recently developed modified embedded atom method (MEAM) potential [52]. Silicon nanobeams and plates were created with initial atomic positions corresponding to the bulk diamond-cubic crystal. The simulation cells were then fully relaxed to a minimum energy state at room temperature and zero pressure. After relaxation, surfaces, such as the one shown in Fig. 12, display a (2×1)–type reconstruction. The details of MDC can be found in [49]. After reaching the equilibrium configurations, the nanoplates were subjected to a quasi-static loading and were again fully relaxed after each strain increment. $\hat{E}$ in extensional mode was defined as

$$\hat{E} = \left(\frac{1}{V_0} \right) \frac{d^2 U}{d\varepsilon^2}, \tag{4}$$

where V_0 is the volume of the nanoplate in fully relaxed zero-strain configuration (undeformed configuration), U is the total energy of the system and ε is the applied strain.

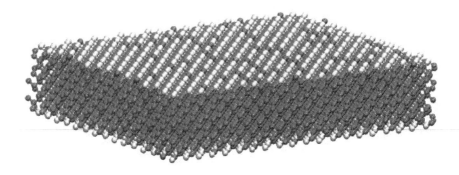

Figure 12. Snapshot of a fully relaxed nanoplate at room temperature. Coloring is according to energy.

The value of $\hat{E}$ was calculated from total energy for simulations with various thicknesses. The results are shown in Fig. 14 (triangles). In order to determine the $\sum S$ using MD, similar to the calculation of $\hat{E}$ described above, the nanoplate is subject to a quasi-static loading and is fully relaxed after each strain increment. The $\sum S$ is extracted in extensional mode as the *2nd*-order derivative of surface energy with respect to the strain, and is about -1 Nm^{-1} [29, 49]. This value was used as an input for Equation 3. Here, it is assumed that the surface elasticity in bending and extensional modes are very close [16].Considering only this surface elasticity effect, a scale-dependence can be calculated and the results are plotted in Fig. 14 as a solid line. The thickness is shown on a logarithmic scale in order to visualize clearly the difference between the experimental results and theoretical investigations. The results of the semi-continuum approach indicate that the surface effects are significant and influence the scale-dependence below 15 nm. It is therefore clear that surface effects can only partially explain the observed scale-dependence.

The native oxide, as an extrinsic surface effect, influences $\hat{E}$ via both its own elastic response and the complex interactions between the oxide and the silicon at the interface [29]. However, it has been shown that the influence of substantial elastic behavior differences between silicon and native oxide is much more significant than the interface elasticity of Si-SiO$_2$. Moreover, during the oxidation process, for every thickness unit of silicon oxide 0.44 units of the silicon surface is consumed, resulting in reducing the thickness of silicon and, consequently, decreasing $\hat{E}$. By taking into account the surface elasticity of the original silicon and the elastic modulus of native oxide layers E_{Ox}, $\hat{E}$ can be estimated as [29]

$$\hat{E} = \frac{E_s t^3 + E_{Ox}\left(1 + \frac{3\Sigma S}{E_b t}\right)\left(8\left(t_{Ox}\right)^3 + 6t^2 t_{Ox} + 12t\left(t_{Ox}\right)^2\right)}{\left(t + 2t_{Ox}\right)^3} \tag{5}$$

where E_{Ox} and t_{Ox} are the native oxide Young's modulus and layers' thickness at the top and bottom of the cantilever. The thickness of the native oxide layers have been reported to vary between 2 to 5 nm [11] and its Young's modulus reported between 50 to 75 GPa [11, 29]. These variations, especially the thickness (since it has a higher influence on the effective elasticity) have been used to calculate the $\hat{E}$ of silicon cantilevers with different thicknesses. The result is shown in Fig. 13. The dotted and dashed lines in Fig. 14 show a comparison between the resultant $\hat{E}$ as a function of cantilever thickness considering the various native oxide scenarios' influence (ranges of thicknesses and Young's modulus reported in the literature) and the measured values. It is clear from the figure that taking the native oxide layer scenarios into account reduces the difference and partially explains the distinctive reduction in $\hat{E}$, yet there is still a considerable gap between experiments and theory.

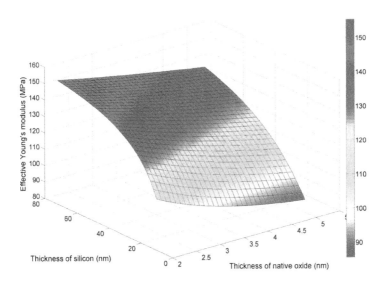

Figure 13. The effective Young's modulus of silicon cantilever as a function of native oxide layers thickness, which varies between 2 to 5 nm and the thickness of the silicon cantilever.

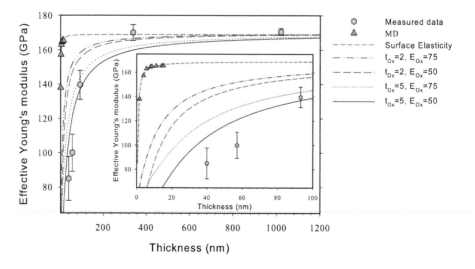

Figure 14. Comparison between experiments and models including surface elasticity and native oxide layers. Hexagons are the measured data via EPI for different thicknesses. The triangles are the results of direct MD calculations in extensional mode. The dashed line illustrates the prediction of scale-dependence when considering surface elasticity via the semi-continuum framework. The dotted-dashed and dotted lines show the prediction when different native oxide scenarios are considered.

One of the most important issues, which has not previously been taken into account, is considering the fact that experimentally tested nanosystems are not defect-free in contrast to the perfect single crystal structures studied using atomistic simulations. In the next section, we discuss the existence of defects and their influence on the scale-dependence.

5. Impact of fabrication-induced defects

There are still some remaining issues that require resolution in order to close the gap between experiments and theory [25]. Among them, accounting for the fact that experimentally fabricated nanosystems are not defect-free, has been recognized as the most important effect, which has not yet been investigated and quantified [25].

As described in Section 2, reducing the cantilever thickness was done by alternating thermal oxidation and etching in HF solution. It was observed that the density of fabrication-induced defects increased when reducing the thickness of the silicon device layer of SOI wafers. During the thermal oxidation process, the lattices of silicon and silicon oxide mismatch, thus introducing stress in both layers [53]. The precipitated oxygen and local strains in Si-SiO_2 interface are likely sources of defects [29]. Further reduction of the thickness of the silicon layer would result in the initiation of defects and cracks, which can extend through the whole layer thickness similar to a pinhole, reaching the underlying BOX layer [53, 29]. These are called HF defects, since the existence of these defects is tested and observed by using the HF etching [38]. The HF solution that reaches the underlying BOX layer etches isotropically the SiO_2 underneath, resulting in a suspended membrane-like layer with a tiny defect in the middle. An optical microscope image of a HF defect at a 14 nm thick silicon device layer is shown in Fig. 15.a. The figure clearly shows a buckled membrane with a HF defect in the middle. The buckling is due to the residual stress between silicon and silicon oxide.

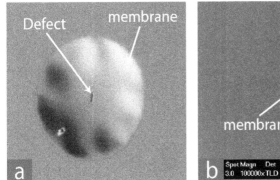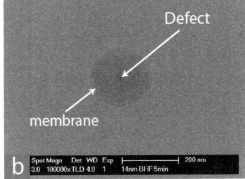

Figure 15. (a) Optical microscope image of a HF defect. (b) SEM of a HF defects. The BOX layer below the HF defects is etched by the HF solution, resulting in the formation of a membrane with a defect in the middle.

The existence of HF defects was also confirmed by SEM, which is observed as a circle (membrane) with the defect at the center (Fig. 15.b). Actually, the HF modifies the defect size and shape, and makes it visible enough to locate it. Using AFM the defect can be observed without HF modifications. The AFM image is shown in Fig. 16.a.

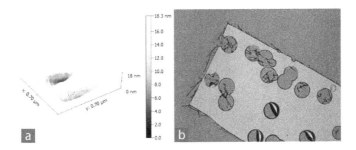

Figure 16. (a) AFM image of a defect without HF modification (shape is caused by a AFM probe). (b) Extending the HF etching of the defects for a longer time causes suspension of relatively large membranes, which finally collapse to the layer below and create a Swiss cheese-like structures.

Continuing the HF etching of the defects results in formation of membranes with a much larger diameter, which are no longer stable and collapse to the layer underneath. This has been shown in Fig. 16.b. What is ultimately remaining, is a structure with big holes, similar to a "Swiss cheese" shape.

The appearance of defects on the surface of the cantilevers can be observed with white light interferometry. As an example, Fig. 17.a shows a white light interferometry image of 40 nm thick silicon cantilevers with defects. The depth profile of one of the defects on the cantilever is shown in Fig. 17.b.

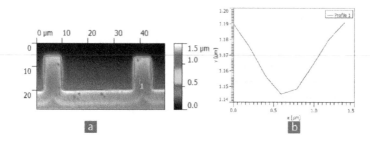

Figure 17. (a) Reconstructed 3D measurements of 40 nm thick silicon cantilevers with defects, using a white light interferometer. (b) The profile measurement of a defect taken along the line 1 in (a), showing the depth and geometry of the defect.

It can be speculated that HF defects contribute to a decrease in $\hat{E}$ and thus, play a role in the gap between experiments and theory of scale-dependence. We utilized MD to examine the extent that the defects contribute to the scale-dependent behavior. Similarly to Section 4, silicon nanoplates were created along the [100] and [110] directions. Defects with cylindrical shapes of different sizes were created by removing the atoms inside the cylinder. Fig. 18 shows the relaxed [100] and [110] nanoplates with defects. After reaching equilibrium configurations, the nanoplates underwent a quasi-static tensile loading. The virial stresses [54] were calculated for each level of strain and $\hat{E}$ was calculated from the slope of the resulting stress-strain curve. Details of the calculations were published in [38]. The MD calculations were repeated for different nanoplates (small, medium and large size) with different dimensions of the defects. The ratio of effective elastic modulus with and without defects, E_d/E_0, versus the ratio of the defects' surface area and total surface area, A_d/A_0 are plotted in Fig. 19. The inset shows the results as a function of the ratio of the number of removed atoms and total number of atoms, N_d/N_T. It is noteworthy that creating the defect did not change the average total energy per atom significantly, except for the atoms near the surface of the defects, which have higher excess energy. An example is shown in Fig. 18.e. The coloring is according to the energy. Apart from the two layers closest to the defect, the energy of the other atoms did not substantially change.

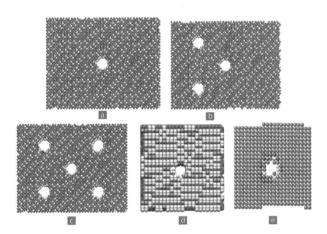

Figure 18. Snapshots of (a) a [100] silicon nanoplate with one defect at the centre, (b) and (c) patterned defects in a [100] silicon nanoplate and (d) [110] nanoplate with one defect at the centre. (e) a mid-plane of a [100] nanoplate with a defect in the centre, indicating the atoms with different energy due to the presence of a defect. Coloring is according to energy.

The reason that the first two layers have a different energy is because removing atoms creates free surfaces inside the defects, causing atoms inside the defect to reconstruct resulting in changes in energy. In order to determine the contribution of each atom to the reduction of $\hat{E}$, the changes in the virial stress of each individual atom as a function of strain, which is the stiffness contribution of each atom, was calculated. The results are shown in Fig. 20, indicating substantial changes when the size of the defects is increased.

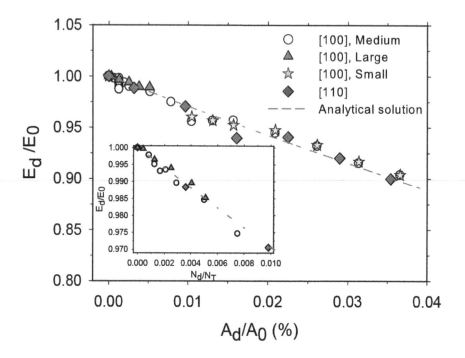

Figure 19. Reduction of the effective Young's modulus versus the size of a defect for different dimensions and directions of nanoplates. The inset shows the re-plotting with respect to the number of removed atoms.

Besides MD, an analytical solution based on continuum theory was used to approximate the effects of defects on $\hat{E}$. It is based on the approximation described in [55] and is as follows:

$$
E_d = \frac{E_0 \left(1 - \dfrac{A_d}{A_0}\right)}{\left(1 + 2\dfrac{A_d}{A_0}\right)}
\tag{6}
$$

The result is plotted in Fig. 19 as a dashed line. A remarkably good agreement between MD and analytical solutions can be seen. The results show that $\hat{E}$ decreases monotonically with increasing defect density (defect volume fraction). The decrease is significant and therefore, it has to be considered as an additional factor contributing to the scale-dependence.

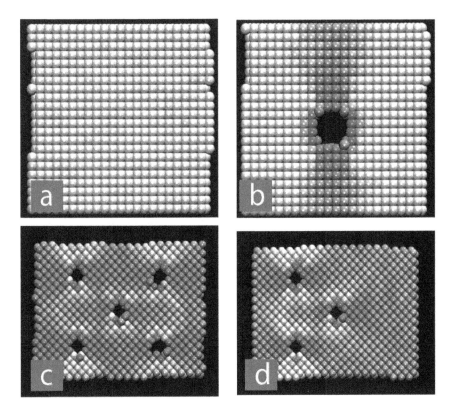

Figure 20. Snapshots of middle layers of silicon (a) [110] nanoplate without defect, (b) [110] nanoplate with a defect in the centre, (c) and (d) [100] nanoplates with patterned defects. Each individual atom is coloured by its stiffness contribution. Blue colours denote atoms of higher stiffness.

The experimental observations, which were described earlier in this section, indicate that the defect density (A_d/A_0) is between 0.005% to 0.12%, depending on the thickness of the silicon device layer [38]. Taking into account the range of experimentally observed defect density and the results presented in Fig. 19, one can include the effect of defects in the scale-dependent $\hat{E}$. The results including the effects of surface elasticity, native oxide layers and fabrication-induced defects were calculated for the thicknesses that were experimentally measured. The results are presented in Fig. 21. Due to variations in the defects' density, an upper and lower limit of the influence of defects is introduced in the figure. The shaded area in the figure demonstrates the extent that defects influence $\hat{E}$. Further quantification of the defects' density versus the thickness of nanocantilevers has to be done. It can be seen, that taking the defects into account, could explain the observed gap between experimental measurements and theoretical calculations.

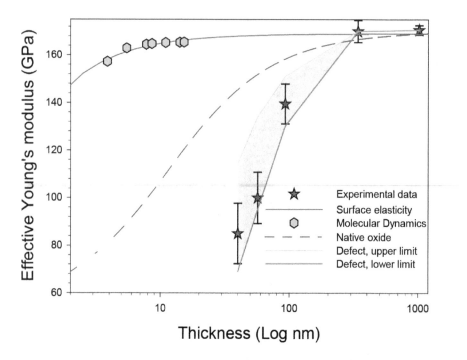

Figure 21. Scale-dependent $\hat{E}$ in bending mode, taking into account the surface elasticity, native oxide layers and the fabrication-induced defects. In order to take into account the variations in the defect density, an upper and a lower limit are introduced. Thus, the scale-dependence observed in silicon nanostructures can be explained by a contribution of surface effects, native oxide layers and fabrication-induced defects.

6. Summary, conclusions and recommendations

The main observations, investigations and discussions to be drawn from the chapter are summarized and concluded in this section.

Silicon based nanodevices are widely used in sensing and actuating applications. For reliable design of such devices, a thorough knowledge of the mechanical properties of these nano-structures is of vital importance. In order to improve the sensitivity, significant research efforts have been directed towards reducing the size of the nanostructures. However, decreasing the size causes the mechanical behavior and the elastic behavior to deviate from the bulk value, known as "scale-dependence" phenomenon.

Two approaches have been explored to study scale-dependent elastic behavior: experimental and theoretical, however a discrepancy exists between the two approaches. The scale at which the scale-dependence starts in experimental measurements is different from that estimated theoretically. Various techniques involving resonance frequency, tensile tests in TEM, atomic force microscopy and nanoindentation have been used to characterize the effective Young's modulus of silicon nanocantilevers and nanowires. Each method involves different assumptions, sources of errors and interpretations. We proposed the use of the electrostatic pull-in instability method in order to avoid most of the issues with other existing methods, as described in Section 3. Using molecular dynamics calculations and the semi-continuum approaches (Section 4) the surface effects on the elastic behavior of silicon nanocantilevers have been investigated. Direct comparison between the surface effects simulations with experimental data from the reliable EPI method showed that although surface effects influence the effective Young's modulus of silicon to some extent, they alone are insufficient to explain the experimental observations.

Another important influence is caused by native oxide layers that exist at the surfaces of the silicon nanocantilevers. The native oxide layers influence the effective Young's modulus of silicon in 3 ways. Through: 1) its distinct elastic response; 2) unknown interactions between the oxide and the silicon at the interface; and, 3) consumption of silicon during oxidation; during oxidation for every unit of silicon oxide 0.44 units of the silicon surface is "consumed". We assumed that the effect of native oxide on the surface elasticity of silicon, or the interface elasticity of Si-SiO2, is not significant compared to the distinct elastic response of the native oxide. Taking the native oxide layers into consideration reduced the difference between experimental measurements and theoretical predictions, yet there was still a considerable difference (Fig. 14).

One of the important issues which was confirmed experimentally, but had not previously been considered in theoretical modeling, was accounting for the fact that experimentally tested nanocantilevers and nanowires are not defect free. Molecular dynamics calculations have been carried out to determine the effects of defects on the effective Young's modulus of silicon nanocantilevers. The conclusion is that the scale effect observed in silicon nanocantilevers can be explained by a contribution of surface effects, native oxide layers and defects. Taking these into account, the gap between the experimental measurements and theory can be closed.

There are a number of issues that are recommended for future research on the size effects:

- The electrostatic pull-in instability method has its own limitations; the application of electrostatic load requires a fairly conductive device and a counter electrode. This implies that the method is only applicable to conductive materials. Moreover, stiction prevention by additional stopper is necessary to ensure the release of the cantilever for multiple measurements. Another fact, that is believed to be minor, but has to be investigated in detail, is the influence of the electric field (electrostatic charges) used in EPI method on the measured value of the effective Young's modulus of nanocantilevers and nanowires.

- Future work on the silicon native oxide layer and its properties will help to explain in more detail the observed size effects. The scale-dependent elastic behavior of silicon native oxide layers has to be investigated. Proper measurements on determination of the thickness and the Young's modulus of silicon native oxide layer as a function of silicon thickness is recommended as further research. In this work, we assumed that the effect of native oxide on the surface elasticity of silicon, or the interface elasticity of Si-SiO$_2$ is not significant. However, a systematic computational study on the interface of Si-SiO$_2$ is essential to determine the extent to which this factor affects the elastic behavior. One of the limiting factors for studying the interface of Si-SiO$_2$ is the lack of a proper potential for Si-SiO$_2$ in MD calculations. Therefore, future research for developing a reliable potential for silicon native oxide and its interface with silicon is also necessary to enable direct comparison with experiments.

In order to better explain the influence of HF defects on the effective Young's modulus of silicon nanosystems, future work on the quantitative determination of the defects and defect density as a function of silicon thickness is necessary. Moreover, the inevitable random distribution of defects, arising from nanocantilever and nanowire synthesis has to be taken into account.

Acknowledgements

This book chapter was financially supported by Enabling Technology program (ETP) for Materials Technology, Dr. L.J.M.G. Dortmans, of Netherlands Organization for Scientific Applied Research (TNO). The authors would like to thank Dr. T.S.D. O'Mahoney and Dr. A. Bossche for their supports, helps and valuable suggestions.

Author details

Hamed Sadeghian[1,2], Fred van Keulen[2] and Hans Goosen[2]

1 Technical Sciences, Netherlands Organization for Applied Scientific Research, TNO, CK, Delft, The Netherlands

2 Structural Optimizations and Mechanics Group, Department of Precision and Microsystems Engineering, Delft University of Technology, CD, Delft, The Netherlands

References

[1] M. A. Hopcroft, W. D. Nix and T. . W. Kenny, "What is the Young's Modulus of Silicon?," *Journal of Microelectromechanical Systems*, vol. 19, no. 2, pp. 229-238, April 2010.

[2] K. R. Virwani, A. P. Malshe, W. F. Schmidt and D. K. Sood, "Young's modulus measurements of silicon nanostructures using a scanning probe system: a non-destructive evaluation approach," *Smart Mater. Struct.*, vol. 12, pp. 1028-1032, 2003.

[3] G. A. Steele, A. K. Huttel, B. Witkamp, M. Poot and H. B. Meerwaldt, "Strong coupling between single electron tunneling and nanomechanical motion," *Science*, vol. 325, no. 5944, pp. 1103-1107, 2009.

[4] H. J. Mamin and D. Rugar, "Sub-attonewton force detection at millikelvin temperatures," *Applied Physics Letters*, vol. 79, no. 20, pp. 3358-3360, 2001.

[5] Naik, O. Buu, M. D. LaHaye, A. D. Armour, A. A. Clerk, M. P. Blencowe and K. C. Schwab, "Cooling a nanomechanical resonator with quantum back action," *Nature*, vol. 443, no. 7108, pp. 193-196, 2006.

[6] X. Li, T. Ono, Y. Wang and M. Esashi, "Ultrathin single-crystalline-silicon cantilever resonators: Fabrication technology and significant specimen size effect on Young's modulus," *Applied Physics Letters*, vol. 83, p. 3081, 2003.

[7] Q. H. Jin, T. Li, Y. L. Wang, X. X. Li, H. Yang and F. F. Xu, "Young's modulus size effect of SCS nanobeam by tensile testing in electron microscopy," in *IEEE Sensors*, 2009.

[8] Q. H. Jin, T. Li, Y. L. Wang, X. L. Gao and F. F. Xu, "Confirmation on the size-dependence of Young's modulus of single crystal silicon from the TEM tensile tests," in *IEEE SENSORS Conference*, 2010.

[9] D.-M. Tang, C.-L. Ren, M.-S. Wang, X. Wei, N. Kawamoto, C. Liu, Y. Bando, M. Mitome, N. Fukata and a. D. Golberg, "Mechanical Properties of Si Nanowires as Revealed by in Situ Transmission Electron Microscopy and Molecular Dynamics Simulations," *Nano Letters*, vol. 12, no. 4, p. 1898–1904, 2012.

[10] K. Asthana, A. Momeni, Y. k. Prasad and R. S.assar, "In situ observation of size-scale effects on the mechanical properties of ZnO nanowires," *Nanotechnology*, vol. 22, p. 265712, 2011.

[11] M. J. Gordon, T. Baron, F. Dhalluin, P. Gentile and P. Ferret, "Size effects in mechanical deformation and fracture of cantilevered silicon nanowires," *Nano Letters*, vol. 9, no. 2, pp. 525-529, 2009.

[12] Y.-S. Sohn, J. Park, G. Yoon, J. Song, S.-W. Jee, J.-H. Lee, S. Na, T. Kwon and a. K. Eom, "Mechanical Properties of Silicon Nanowires," *Nanoscale Research Letters*, vol. 5, no. 1, pp. 211-216, 2010.

[13] B. Lee and R. E. Rudd, "First-principles calculation of mechanical properties of s¡ 001 nanowires and comparison to nanomechanical theory," *Physical Review B*, vol. 75, no. 19, p. 195328, 2007.

[14] Y. Umeno, A. Kushima, T. Kitamura, P. Gumbsch and J. Li, "Ab initio study of the surface properties and ideal strength of (100) silicon thin films," *physical Review B*, vol. 72, no. 16, p. 165431, 2005.

[15] B. Lee and R. E. Rudd, "First-principles study of the young's modulus of si 001 nano-wires.," *Physical Review B*, vol. 75, no. 4, p. 041305, 2007.

[16] M. T. McDowell, A. M. Leach and K. Gall, "Bending and tensile deformation of met-allic nanowires," *Modelling and Simulation in Materials Science and engineering*, vol. 16, no. 4, p. 045003, 2008.

[17] S. H. Woo, L. G. Zhou, H. Hanchen and S. C. Timothy, "Nanoplate elasticity under surface reconstruction," *Applied Physics Letters*, vol. 86, no. 15, p. 151912, 2005.

[18] S. H. Park, J. S. Kim, J. H. Park, J. S. Lee, Y. K. Choi and a. O. M. Kwon, "Molecular dynamics study on size-dependent elastic properties of silicon nanocantilevers," *Thin Solid Films*, vol. 492, no. 1-2, pp. 285-289, 2005.

[19] K. Kang and W. Cai, "Brittle and ductile fracture of semiconductor nanowires: a mo-lecular dynamics simulations," *Philosophical Magazine*, vol. 87, no. 14, pp. 2169-2189, 2007.

[20] R. E. Miller and V. B. Shenoy, "Size-dependent elastic properties of nanosized struc-tural elements," *Nanotechnology*, vol. 11, no. 3, pp. 139-147, 2000.

[21] G. Wei, Y. Shouwen and H. Ganyun, "Finite element characterization of the size-de-pendent mechanical behaviour in nanosystems," *Nanotechnology*, vol. 17, no. 4, pp. 1118-11122, 2006.

[22] J. Wang, Q. A. Huang and H. Yu, "Size and temperature dependence of young's modulus of a silicon nano-plate," *Journal of Physics D: Applied physics*, vol. 41, no. 16, p. 165406, 2008.

[23] H. Sadeghian, J. F. L. Goosen, A. Bossche and F. v. Keulen, "Surface stress-induced change in overall elastic behavior and self-bending of ultra-thin cantilever plates," *Applied Physics Letters*, vol. 94, p. 231908, 2009.

[24] R. Agrawal and H. D. Espinosa, "Multiscale experiments: State of the art and remain-ing challenges," *Journal of Engineering Materials and Technology*, vol. 131, no. 4, p. 041208, 2009.

[25] H. S. Park, W. Cai, H. D. Espinosa and H. Huang, "Mechanics of crystalline nano-wires," *MRS BULLETIN*, vol. 34, pp. 178-183, 2009.

[26] X. Han, K. Zheng, Y. Zhang, X. Zhang, Z. Zhang and Z. Wang, "Low-temperature in-situ large-strain plasticity of silicon nanowires," *Advanced Materials*, vol. 19, no. 16, pp. 2112-2118, 2007.

[27] H. S. Park and P. A. Klein, "Surface Cauchy-Born analysis of surface stress effects on metallic nanowires," *PHYSICAL REVIEW B*, vol. 75, p. 085408, 2007.

[28] G. Yun and H. S. Park, "Surface stress effects on the bending properties of fcc metal nanowires," *Phys. Rev. B*, vol. 79, p. 195421, 2009.

[29] H. Sadeghian, C.-K. Yang, J. F. L. Goosen, A. Bossche, U. Staufer, P. J. French and F. v. Keulen, "Effects of size and defects on the elasticity of silicon nanocantilevers," *Journal of Micromechanics and Microengineering,* vol. 20, p. 064012, 2010.

[30] D. Zeng, "Tunable resonant frequencies for determining Young's moduli of nano-wires," *Journal of Applied Physics,* vol. 105, no. 11, pp. 114311 - 114311-4 , 2009.

[31] M. T. McDowell, A. M. Leach and a. K. Gall, "On the elastic modulus of metallic nanowires," *Nano Letters,* vol. 8, no. 11, pp. 3613-3618, 2008.

[32] H. Sadeghian, C. K. Yang, J. F. L. Goosen, E. v. d. Drift, A. Bossche, P. J. French and F. v. Keulen, "Characterizing size-dependent effective elastic modulus of silicon nano-cantilevers using electrostatic pull-in instability," *Applied Physics Letters,* vol. 94, no. 22, p. 221903, 2009.

[33] H. Sadeghian, H. Goosen, A. Bossche and F. v. Keulen, "Application of electrostatic pull-in instability on sensing adsorbate stiffness in nanomechanical resonators," *Thin Solid Films,* vol. 518, pp. 5018-5021, 2010.

[34] H. Sadeghian, C. K. Yang, J. F. L. Goosen, A. Bossche, P. J. French and F. v. Keulen, "Quantitative analysis and decoupling of mass and stiffness effects in cantilever mass sensors," in *Sensors, 2010 IEEE,* 2010.

[35] R. Dingreville, A. J. Kulkarni, M. Zhou and a. J. Qu, "A semi-analytical method for quantifying the size-dependent elasticity of nanostructures," *Modelling and Simulation in Materials Science and Engineering,* vol. 16, no. 2, p. 025002, 2008.

[36] M. J. Lachut and J. E. Sader, "Effect of surface stress on the stiffness of cantilever plates," *Physical Review Letters,* vol. 99, no. 20, p. 206102, 2007.

[37] K. S. Hwang, K. Eom, J. H. Lee, D. W. Chun, B. H. Cha, D. S. Yoon, T. S Kim and J. H. Park, "Dominant surface stress driven by biomolecular interactions in the dynamical response of nanomechanical microcantilevers," *Applied Physics Letters,* vol. 89, no. 17, p. 173905, 2006.

[38] H. Sadeghian, H. Goosen, A. Bossche, B. Thijsse and F. v. Keulen1, "On the size-de-pendent elasticity of silicon nanocantilevers: impact of defects," *Journal of Physics D: Applied Physics,* vol. 44, p. 072001, 2011.

[39] H. Sadeghian, C.-K. Yang, K. B. Gavan, J. F. L. Goosen, E. W. J. M. v. d. Drift, H. S. J. v. d. Zant, A. Bossche, P. J. French and F. v. Keulen, "Some considerations of effects-induced errors in resonant cantilevers with the laser deflection method," *Journal of Micromechanics and Microengineering,* vol. 20, no. 10, p. 105027, 2010.

[40] J. Jung, D. Seo, G. Park, S. Ryu and H. Song, "Ag-Au-Ag Heterometal Nanowires: Synthesis, Diameter Control, and Dual Transversal," *J. Phys. Chem. C.,* vol. 114, p. 12529–12534, 2010.

[41] R. Agrawal, B. Peng, E. E. Gdoutos and H. D. Espinosa, "Elasticity Size Effects in ZnO Nanowires–A Combined Experimental-Computational Approach," *Nano Letters,* vol. 8, no. 11, p. 3668–3674, 2008.

[42] U. Bhaskara, V. Passia, S. Houria, E. Escobedo-Cousina, S. H. Olsena, T. Pardoena and J.-P. Raskina, "On-chip tensile testing of nanoscale silicon free-standing beams," *Journal of Materials Research,* vol. 27, no. 03, pp. 571-579, 2011.

[43] Y. Xia, P. Yang, Y. Sun, Y. Wu, B. Mayers, B. Gates, Y. Yin, F. Kim and H. Yan, "One-Dimensional Nanostructures: Synthesis, Characterization, and Applications," *Advanced Materials,* vol. 15, no. 5, pp. 353-389, 2003.

[44] H. Sadeghian, G. Rezazadeh and P. M. Osterberg, "Application of the generalized differential quadrature method to the study of pull-in phenomena of MEMS switches," *IEEE/ASME Journal of Microelectromechanical Systems,* vol. 16, no. 6, pp. 1334-1340, 2007.

[45] O. Furmanchuk, O. Isayev, T. C. Dinadayalane, D. Leszczynska and J. Leszczynski, "Mechanical properties of silicon nanowires," *WIREs Comput Mol Sci,* vol. 00, 2012.

[46] W. McFarland, M. A. Poggi, L. A. Bottomley and J. S. Colton, "Characterization of microcantilevers solely by frequency response acquisition," *Journal of Micromechanics and Microengineering,* vol. 15, no. 4, p. 785, 2005.

[47] G.-F. Wang and X.-Q. Feng, "Effects of surface elasticity and residual surface tension on the natural frequency of microbeams," *Applied Physics Letters,* vol. 90, no. 23, p. 231904, 2007.

[48] F. H. Streitz, R. C. Cammarata and K. Sieradzki, "Surface-stress effects on elastic properties. i. thin metal films," *Physical Review B,* vol. 49, no. 15, p. 10699–10706, 1994.

[49] H. Sadeghian, J. F. Goosen, A. Bossche, B. J. Thijsse and F. v. Keulen, "Effects of size and surface on the elasticity of silicon nanoplates: Molecular dynamics and semi-continuum approaches," *Thin Solid Films,* vol. 520, no. 1, pp. 391-399, 2011.

[50] J. Zang, M. Huang and F. Liu, "Mechanism for Nanotube Formation from Self-Bending Nanofilms Driven by Atomic-Scale Surface-Stress Imbalance," *PHYSICAL REVIEW LETTERS,* vol. 98, p. 146102, 2007.

[51] J. Zang and F. Liu, "Theory of bending of Si nanocantilevers induced by molecular adsorption: a modified Stoney formula for the calibration of nanomechanochemical sensors," *Nanotechnology,* vol. 18, p. 405501, 2007.

[52] M. Timonova, B.-J. Lee and B. J.thijsse, "Sputter erosion of Si(0 0 1) using a new silicon MEAM potential and different thermostats," *Nuclear Instruments and Methods in Physics Research Section B,* vol. 255, no. 1, p. 195–201, 2007.

[53] O. Naumova, E. Vohmina, T. Gavrilova, N. Dudchenko, D. Nikolaev, E. Spesivtsev
      and V. Popov, "Properties of silicon nanolayers on insulator," *Materials Science and
      Engineering: B,* vol. 135, no. 3, pp. 238-241, 2006.

[54] D. H. Tsai, "The virial theorem and stress calculation in molecular dynamics," *J.
      Chem. Phys.,* vol. 70, pp. 1375-82, 1979.

[55] Y. Shevlyakov and A. Skoblin, "Relative stiffness of irregularly perforated plates," *J.
      Math. Sci,* vol. 65, pp. 1389-95, 1993.

Chapter 7

Laser Micromachining for Gallium Nitride Based Light-Emitting Diodes

Kwun Nam Hui and Kwan San Hui

Additional information is available at the end of the chapter

1. Introduction

Micromachining with nanosecond laser pulses is a powerful tool that is suitable for replacing or complementing traditional wafer processes, such as dicing and etching, as well as advanced process developments, such as laser lift-off [1], laser-assisted machining [2] and medical and biotechnology research [3]. Tightly-focused nano-second laser pulses can enable micromachining with much higher precision and dimensions down to several micrometers [4]. For more advanced applications, micromachining parameters, such as laser wavelength, pulse energy, repetition rate and pulse duration, should be considered seriously. Drilling and cutting with nanosecond, or even femtosecond ultraviolet (UV) laser pulses has been reported to produce very small heat-affected zones (HAZ) [5]. Recently, laser micromachining is being adopted gradually for gallium nitride (GaN)-based light-emitting diodes (LEDs). Because epitaxial GaN layers are typically grown on sapphire, the separation of fabricated LED dies is commonly achieved by wafer sawing, which is slow and expensive. The use of high energy laser pulses increases the process efficiency and enables a high packing density of chips through the reduced dimensions of the scribe lanes.

One of the typical configurations of laser micromachining relies on the laser scanner head, which steers the beam into the incident direction. Although it is relatively fast, mechanical vibrations tend to be magnified, resulting in a loss of pattern resolution. Alternatively, the sample to be micro-machined can be mounted onto a precision motorized translation module while the optical beam remains static, which is more suitable for processing the optical microstructures. In a conventional laser micro-machining setup for wafer dicing, the focused laser beam is incident perpendicular to the sample to be processed, so that only two dimensional patterns can be generated and vertical cuts can be obtained. Projecting the beam at an oblique angle to the sample enables three-dimensional micromachining. Nevertheless, it cannot mount the translation module at a tilted angle because it will result in severe beam distortion. In the proposed approach, a laser beam turning mirror was introduced to the optical path to achieve a continuously-tunable range of tilting angles for beam projection, while retaining the beam quality. Laser micromachining is a potential simple, inexpensive and high-throughput alternative method for creating geometrically-shaped GaN LEDs compared to other available technologies [6,7,8]. K.N. Hui et al. reported the effectiveness of laser micro-machining incorporated with GaN semiconductors to achieve high light extraction GaN LEDs [7,9] and color tunable vertically-stacked LEDs in solid-state lighting applications [7,10].

This chapter examines the experimental process of laser micromachining, and the structural and optical properties of laser micromachining LED chips with a range of geometries. The optical characterization of LED, particularly the light extraction efficiency of geometrically-shaped LEDs, is discussed because the light extraction efficiency plays an important role in achieving high luminous efficacy LEDs. Finally, several applications derived from the utilization of laser micromachining, e.g. geometrically-shaped LED, angularly uniform white LEDs, and vertically-stacked polychromatic LEDs are presented.

2. Process of laser micromachining

One attractive feature of laser processes is its ability to remove material. Termed laser ablation, material removal can be achieved by physical or chemical microscopic mechanisms. Because lasers can be focused on a small spot with high energy density, precision machining of the features on the micrometer or tens of micrometer scale is possible. For example, E. Gu et al. examined the drilling of holes and micro-trenches in a free-standing GaN substrate by pulsed UV laser ablation[4]. Another use of laser ablation in LED industries is wafer dicing. As GaN is normally grown on sapphire, and sapphire is the second hardest material in the world, a diamond blade is the only viable tool for mechanical dicing. On the other hand, a diamond blade often deviates from its intended dicing direction when the blades are thin, causing chipping or even device damage. With laser dicing, the dicing path can be controlled with high precision. In addition, the spacing between the individual LED dies can be reduced to a size comparable to the laser spot size, leading to an increase in die density.

Simple laser micromachining consists of a UV laser source, beam focusing optics and an x-y motorized translation stage, as shown in Figure 1. The laser source is a third harmonic ND:YLF diode-pumped solid state (DPSS) laser manufactured by Spectra Physics. The laser emits at 349 nm, and the pulse repetition rate ranges from single pulse to 5 kHz. At a reference diode current of 3.2 A, the pulse energy is 120 µJ at a repetition rate of 1 kHz, with a pulse width of approximately 4 nanoseconds. The TEM00 beam allows for tight focusing, offering a high spatial resolution. After beam expansion and collimation using a beam expander, the laser beam is reflected 90° using a dielectric laser line mirror and is focused onto the horizontal machining plane to a very tiny spot, several micrometers in diameter, with a focusing triplet. All optics used are made from UV-fused silica and are anti-reflection (AR) coated. The additional feature of this set-up, as illustrated in the schematic diagram of Figure 1, is the insertion of a UV mirror at an oblique angle within the optical path between the focusing optics and machining plane, which deflects the convergent beam to strike the sample at an oblique angle to the horizontal working plane. The size of the beam at the focal point is not only limited by the capability of the UV objective lens but is also sensitive to the coaxiality of the optics. With this modified set-up, it is relatively easy to optimize and monitor the beam through the tube lens imaged with a CCD camera. Once the optical setup is optimized before inserting the tilting mirror, the mirror can be inserted without affecting the coaxiality of the laser beam, so that the dimensions of the beam spot are unaffected.

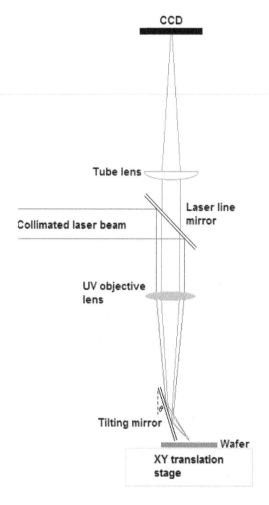

Figure 1. Experimental setup of laser micro-machining.

The angle of incidence of the deflected laser beam on the wafer is 2θ, where θ, as indicated in Figure 1, is the angle between the plane of the mirror and the normal. This angle can be precisely adjusted by mounting the mirror onto a rotation stage. Therefore, the incident angle can be varied over a wide range. In this experiment, a UV objective with a focal length of 75 mm was used based on two considerations. First, the focal length should be long enough to accommodate the mirror in the optical path. Secondly, an ideal tool for the fabrication of microstructures should have a very long penetration depth and negligible lateral dispersion. Nevertheless, an objective lens with a longer focal length also produces a larger focused beam spot. The two parameters are related by the following equation:

$$d = \frac{4\lambda M^2 f}{\pi D} \tag{1}$$

where M^2 quantifies the beam quality, λ is the wavelength of the laser beam, f is the focal length and D is the diameter of the incident beam.

3. Characterization of laser micromachining

3.1. Depth of micro-trenches as a function of the scan cycles

After laser micromachining, the depth of the micro-trenches pattern can be observed by scanning electron microscopy (SEM) to check the effectiveness of the approach. The quality of the cleave can be quantified by the width, depth, linearity and sidewall roughness of the trench formed by the laser beam. Because the focal length of the focusing lens (f = 75 mm) is much longer than the thickness of the GaN layer on the sapphire wafer (t = 420 μm), the depth of the trench depends mainly on the number of micromachining cycles. The number of cycles is controlled by configuring the translation stage to repeat its linear path several times. As the position reproducibility of the stage is better than 5 μm, increasing the number of cycles should not contribute significantly to the width of the feature. X.H. Wang et al. [11] reported the cross-sectional optical image of a GaN layer on a 420 μm thick sapphire wafer that had been micro-machined with an incident beam inclined at 45° with scan cycles ranging from 1 to 10. These incisions were carried out by setting the laser pulse energy to 54 μJ at a repetition rate of 2 kHz. Figure 2 shows the relationship between the inclined cutting depth and the number of passes of the beam. After the first pass of the beam, a narrow trench with a width of ~20 μm and depth of ~220 μm was formed. Successive scans of the beam along the trench resulted in further deepening and widening but the extent was increased at a decreasing rate. The depth of the trench depends on the effective penetration of the beam. From the second scan onwards, the beam needs to pass through the narrow gap before reaching the bottom of the trench for further machining. The energy available at this point was attenuated, which is partly due to lateral machining of the channel (causing undesirable widening), absorption and diffraction effects. Therefore, the depth of the trench tends to saturate after multiple scans.

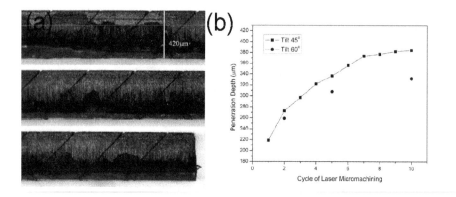

Figure 2. (a) Cross-sectional optical micrograph of laser micro-machined micro-trenches at an inclination angle of 45° at a range of scan cycles of between 1 and 10 (left to right then down), and (b) depth of tiling micro-trenches as a function of the scan cycles.

3.2. Topography of laser micromachining

In addition to the scan speed and number of scan cycles, the focus offset and pulse energy are two important parameters controlling the quality and topography of the micro-trenches of GaN structures. The focus offset level is defined as the distance shifted away from the focal plane; the downward direction is positive. Y.H. Mak et al. [12] showed that micro-trenches with different topographies can be obtained precisely by controlling the focus offset and pulse energy of the laser beam. For example, in Figure 3(a), the sample is positioned near the focal plane (300 µm from focal plane); the laser beam ablates both the GaN and sapphire layers. A V-shaped valley is formed in the sapphire layer due to the Gaussian beam shape. At the optimal focal offset plane of 450 µm, as shown in Figure 3(b), ablation terminates automatically at the GaN/sapphire interface because the laser fluence decreases below the ablation threshold value for sapphire, resulting in the exposure of a flat and smooth sapphire bottom surface. At a larger focus offset plane of 600 µm, the GaN layer is not removed completely, leaving a shallow and rugged trench on the surface (Figure 3(c)).

Figure 3(d-f) illustrates the laser micromachined micro-trenches formed at three different pulse energies (45, 23, and 7 µJ) with the other parameters constant, and the focus offset is kept at the optimal value of 450 µm. When the pulse energy is set to 45 µJ, GaN and sapphire are ablated to form a V-shaped trench (Figure 3(d)), which is similar to that with a smaller focus offset. On the other hand, a low pulse energy results in shallow micro-trenches, which is similar to the large focus offset.

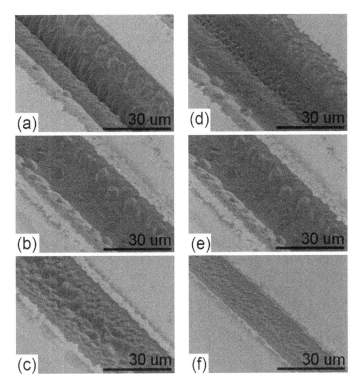

Figure 3. SEM images of micro-trenches formed by laser micromachining at different focus offset planes (with the pulse energy, pulse repetition rate and scan speed fixed at 23 µJ, 1 kHz, and 25 µm/s, respectively.): (a) small offset of 300 µm; (b) optimal offset of 450 µm; (c) large offset of 600 µm, and (d) pulse energy of 45µJ, (e) 23 µJ, and (f) 7 µJ (with the focus offset level, pulse repetition rate, and scan speed are fixed at 450 µm, 1 kHz, and 25 µm/s, respectively.)

4. Laser micromachining applications

4.1. Geometrical shaped LED

The development of LEDs with high optical output power has been the driving force of next generation solid-state lighting [13]. On the other hand, the optical output power of the-state-of-art LEDs is still insufficient for making them practically viable. The large refractive index difference between nitride material (η_{GaN} = 2.585) and air (η_{Air} = 1), giving rise to a total internal reflection at the interfaces, is the major cause for the lower-than-expected light extraction efficiency. In addition, conventional LED chips with a cuboid geometry and a Lambertian emission pattern often have a light extraction efficiency of < 20%. Several methods have been proposed to alleviate these issues, such as flip-chip LEDs [14], photonic crystals [15] and surface texturing [16]. These proposed methods, however, are energy consuming, low throughput, and often utilize expensive equipment, highlighting the need to search for alternative low cost methods that can be fully adopted in industrial mass production and enhance the light output intensity of LEDs significantly. Recently, the effect of geometrical chip-shaping realized with laser micromachining or other methods is being gradually recognized as a promising alternative technique for optimizing the efficiency and for modulating the emission pattern [17]. W.F. Fu et al. [8] reported that the geometrical shaping of LEDs by laser micromachining is an effective approach for enhancing the light extraction efficiency of a conventional cuboid LED (inclination angle of 90°) of 18.3% up to 33.9% in truncated pyramidal (inclination angle of 50°) LED geometry. This approach offers significant increases in light extraction efficiency of up to 85.2%, which is the highest value reported thus far.

Figure 4(a) presents the mechanism of enhanced light extraction with tiled sidewalls, showing the additional light extraction channel from the top surface as well as from the sidewalls due to reflections on the tilted sidewalls. According to the ray-tracing simulation, the light extraction efficiency depends on the inclination angle. Figure 4(b) shows the light extraction efficiency as a function of the inclination angle. Figure 4(c) shows a SEM image of an InGaN LED die with a truncated pyramidal geometry (TP-LED) fabricated by laser micromachining. Figure 5(a) and 5(b) show the operation images of a cuboid LED and a TP-LED, respectively, and Figure 5(c) presents their light output–current (L-I) characteristics. At lower driving currents (50 mA), the average light enhancement factor was 88.6%, which is consistent with the theoretical prediction of 85.2%. Such significant improvement in light extraction efficiency highlights the effectiveness of geometrical chip-shaping, particularly with the present approach based on laser micromachining.

4.2. Angularly uniform white LED

White LEDs are used widely in commercial applications, such as solid-state lighting, liquid crystal display (LCD) backlighting and signaling, owing to their energy efficiency and mercury-free composition. Currently, color down-conversion and color-mixing are the two mainstream methods of producing white LEDs. The use of phosphors as a conversion agent is used widely in commercial products. Nevertheless, the limited conversion efficiency from shorter wavelengths (typically blue at approximately 470 nm) to a longer wavelengths means the benefits of LEDs can never be fully achieved. Placing three LEDs (red, green and blue) into a single package (the RGB approach) resolves this deficiency but introduces severe issues with color uniformity and homogeneity. For phosphor-coated LEDs, the placement and method of the phosphor coating will also affect the color uniformity considerably, whereby emission homogeneity is an important attribute for many applications. For example, the phosphor coating process went through a reflow process to cover both the top and sidewalls, resulting in a non-uniform distribution and coating thickness, particularly at the edge of the chip. Such coating thickness non-uniformity, coupled with the unequal light emission from the top and sidewall, results in a non-uniformity of color emission from different viewing angles [18]. Therefore, tailoring the light emission pattern from the point of view of geometrically-shaped LED plays an important role in achieving high angular color uniformity. L. Zhu et al. [19] found that a truncated cone (TC) structure integrating an Al mirror reflectors on the sidewall and bottom surfaces is an effective approach for improving the angular color uniformity of white LEDs, showing 37% enhancement in color uniformity, compared to the conventional cuboid structure. Figure 6 presents three different coating profiles, phosphor-slurry coating, conformal coating and remote phosphor coating, along with the proposed white LED employing a TC structure.

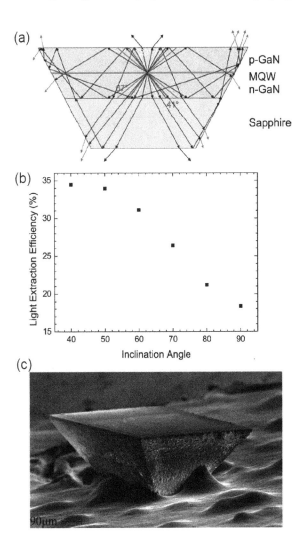

Figure 4. a) Schematic diagram showing an exemplary light ray within a TP-LED. The arrows in blue indicate the additional extracted rays due to the inclined sidewalls. (b) Light extraction efficiency of an inclined sidewall LED as a function of the inclination angle of the sidewall. (c) SEM image of an InGaN LED die with a truncated pyramidal geometry.

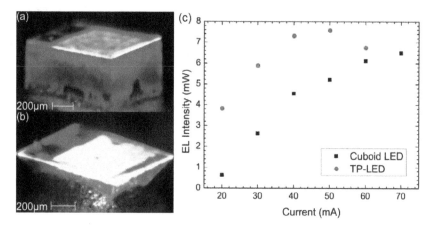

Figure 5. Optical micrographs showing cuboid LED (a) and TP-LED (b) biased at 10 mA, and (c) L-I curve comparing the performances of the TP-LED and cuboid LED.

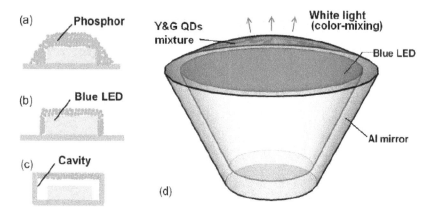

Figure 6. Schematic diagram showing the cross-sectional views of LED coating using three different methods: (a) phosphor-slurry coating; (b) conformal coating; (c) remote phosphor coating; and (d) 3D schematic diagram of a TC-LED.

Figure 7 shows operational images of a TC-LED, together with the reference LED (circular LED without TC structure). Quantum dots (QDs) consisting of green (540 nm) and yellow (560 nm) light-emitting QDs (Evident Technologies) are used as color-conversion agents, which are mixed together to a 7 : 5 volume ratio for balanced white light emission. The mixture is then blended with a transparent UV epoxy (Norland 61), following which a small volume of the slurry is dispensed onto the chips. The chips are packaged into standard TO-cans. To minimize these effects, the epoxy is spin-coated onto the chips to ensure evenness of the coating. The completed devices were tested at a bias current of

20 mA. Figure 7(c) and 7(e) shows the emission from the reference LED structure from two different angles. A ring of yellow light is clearly observed at the periphery of the circular chip, which was attributed to the relatively thicker coating on the sidewalls. This effect is suppressed with the TC-LED structure, as shown in Figure 7(d) and 7(f), where uniform white light emission can be observed from different angles.

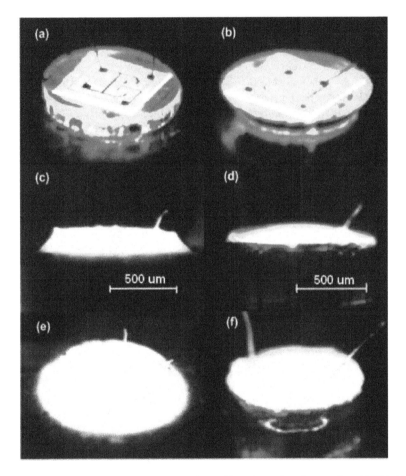

Figure 7. Operational images of the packaged devices: (a) and (b) are the circular reference LED and TC-LED; (c) and (e) show the QD-coated circular reference LED, (d) and (f) show the QD-coated TC-LED.

4.3. Vertically-stacked polychromatic LED

Recent advances in the performance of high power LEDs have led to the development of novel illumination sources with added functionality and intelligence. These advanced technologies have attracted attention from both academia and industry due to the increasing demand for dynamic lighting not only for outdoor activities, such as accent and task lighting, stage and studio, but also in indoor activities, such as tunable interior mood lighting [20]. Numerous spectral conversion schemes have been adopted for polychromatic LEDs, such as phosphors [21], polymer dyes [22] and CdSe/CdS quantum dots [23] for color down-conversion. These schemes, however, inevitably suffer from energy loss due to Stokes shift in the conversion processes, as well as scattering losses associated with the particles. Another conversion-free approach involves the combination of three discrete LED chips of the primary colors red (R), green (G) and blue (B) arranged side-by-side on the same plane to generate a polychromatic spectrum. On the other hand, challenges exist in uniform color mixing, both spatially and angularly in such configurations [24]. K.N. Hui et al. [7,10] pioneered the stacking of TP-LED chips in that instead of placing the RGB chips onto the same plane, the individual LEDs were arranged in a vertically-stacked configuration to produce LED-stacks. The LED-stacks consisted of an InGaN/GaN blue LED (470 nm) stacked onto an InGaN/GaN green LED (520 nm), which is subsequently stacked onto an AlGaInP/GaAs red LED (650 nm). Such a stacking strategy ensures optimal color mixing and minimal absorption losses. Figure 8 provides a schematic diagram of the proposed device.

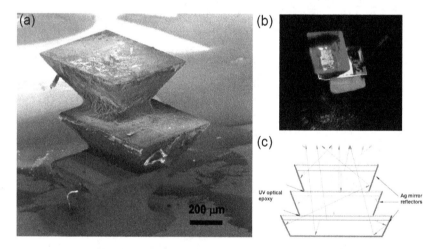

Figure 8. (a) SEM image of an LED-stack assembled from LED chips with a truncated pyramid geometry, (b) operational image of the LED-stacks, and (c) schematic diagram showing the mixing of light inside the LED-stacks.

Figure 9 shows the corresponding electroluminescence (EL) spectral data and optical emission images of the LED-stacks driven at the tested voltage combinations. A wide range of colors can be obtained from the stacked LEDs, which depend on the choice of the individual LED chip with a specific wavelength and spectral bandwidth. The proposed LED-stacks with a color tunable function have potential applications in high-resolution panel displays. Table 1 summarizes the combinations of an applied driving current for generating a range of colors. In this study, the overall performance of the packaged stacked LEDs device was measured in a calibrated 12-inch integrating sphere. The optical signal was channeled using an optical fiber to an optical spectrometer. At a total driving current of 20 mA, the LED-stacks produced a luminous efficacy of 33 lm/W, whereas the commercial RGB LEDs produced a luminous efficacy of 30 lm/W. The corresponding CIE coordinates, CRI and CCT values of the LED-stacks were (0.32, 0.33), 69, and 6300 K, respectively, which is a promising result for a prototype device.

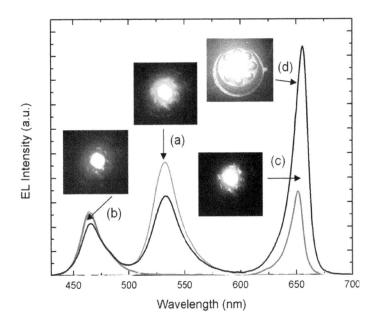

Figure 9. (a)-(d) illustrates the electroluminescence spectrum of the various colors emitted by the LED-stacks. The inserted images show the corresponding devices.

Curve	Red LED	Green LED	Blue LED	CIE (x,y)
(a)		2.57 V (2 mA)	2.70 V (10 mA)	(0.18, 0.27)
(b)	2.60 V (2 mA)		2.53 V (2 mA)	(0.31, 0.13)
(c)	2.97 V (5 mA)	2.61 V (2 mA)		(0.46, 0.53)
(d)	3.72 V (14 mA)	2.68 V (4 mA)	2.53 V (2 mA)	(0.32, 0.33)

Table 1. Biased voltage (current) of the electroluminescence spectrum for Figure 9(a) to (d).

Figure 10 shows the CIE chromaticity as a function of the viewing angle of the LED-stacks and commercial RGB LED. The CIE coordinates of light emission collected at the normal incidence (0°) from the top of the LED-stacks and the commercial RGB LEDs were

(0.32, 0.33) and (0.33, 0.33), respectively. As the angle of observation was increased from 0° to 70 in 10° increments, shifting of the CIE coordinate from the stacked LEDs was insignificant and the device emitted mixed white light. On the other hand, under the same test, the CIE coordinate of the commercial RGB LED shifted to a yellowish white. At an observation angle of 70°, the farthest shift of the CIE coordinate away from the initial CIE coordinates of the stacked LEDs was (0.29, 0.29), whereas the CIE coordinate of the commercial RGB LED were shifted to (0.41, 0.39). These results highlight the effectiveness of the vertical stacking of LEDs in achieving uniform color mixing.

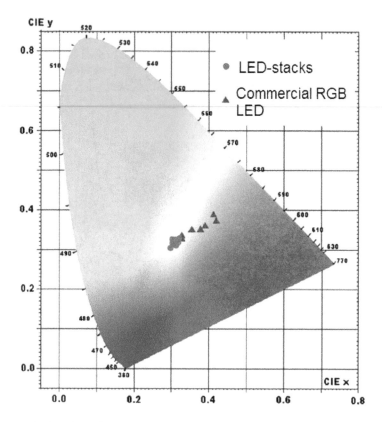

Figure 10. Angular dependence of the CIE chromaticity on the LED-stacks and commercial RGB LED.

5. Summary

This chapter reviewed several studies of laser micromachining on GaN LEDs. Based on the knowledge of several key parameters, such as the scan cycle, scan speed, pulse energy, and offset focus of laser micromachining, laser micromachining is a feasible approach for obtaining high quality and performance GaN LEDs. On the other hand, a few laser micromachining applications have been addressed. Geometrically-shaped LEDs provides an effective way of enhancing the light extraction efficiency of LEDs. Angularly uniform white LEDs help improve the angular color uniformity of white LEDs. Vertically-stacked polychromatic LEDs can improve light extraction and has potential applications including high uniformity color-tunable light sources and conversion-free white LED. The mass production of high light extraction efficiency LEDs, high angular color uniformity white LEDs and high functionality GaN-based LEDs may be realized in the near future when the laser micromachining approach is adopted widely.

Acknowledgements

This work was carried out at the University of Hong Kong, City University of Hong Kong and Pusan National University. The authors would like to thank H.W. Choi for the opportunity to work on laser micromachining for GaN LEDs. This work was supported by Basic Science Research Program through the National Research Foundation of Korea (NRF) funded by the Ministry of Education, Science and Technology (2010-0023418).

Author details

Kwun Nam Hui[1*] and Kwan San Hui[2,3]

*Address all correspondence to: bizhui@pusan.ac.kr; kwanshui@um.cityu.edu.hk

1 Department of Materials Science and Engineering, Pusan National University, Geumjeong-Gu, Busan, Republic of Korea

2 Department of Systems Engineering & Engineering Management, City University of Hong Kong, Kowloon Tong, Hong Kong, China

3 Department of Mechanical Engineering, Hanyang University, Seongdong-gu, Seoul, Republic of Korea

References

[1] A. Elgawadi, J. Krasinski, G. Gainer et al., "Modification of the anomalous optical transitions in multilayer AlGaN-based nanoheterostructure using a nonbonding laser lift-off technique," J Appl Phys 103 (12), 123512 (2008).

[2] Y. G. Tian and Y. C. Shin, "Laser-assisted machining of damage-free silicon nitride parts with complex geometric features via in-process control of laser power," J Am Ceram Soc 89 (11), 3397-3405 (2006).

[3] N. Muhammad, D. Whitehead, A. Boor et al., "Picosecond laser micromachining of nitinol and platinum-iridium alloy for coronary stent applications," Appl Phys a-Mater 106 (3), 607-617 (2012)

[4] E. Gu, C. W. Jeon, H. W. Choi et al., "Micromachining and dicing of sapphire, gallium nitride and micro LED devices with UV copper vapour laser," Thin Solid Films 453, 462-466 (2004).

[5] P. Molian, B. Pecholt, and S. Gupta, "Picosecond pulsed laser ablation and micromachining of 4H-SiC wafers," Appl Surf Sci 255 (8), 4515-4520 (2009).

[6] G. Y. Mak, E. Y. Lam, and H. W. Choi, "Interconnected alternating-current light-emitting diode arrays isolated by laser micromachining," J Vac Sci Technol B 29 (1) (2011)

[7] K. N. Hui, X. H. Wang, Z. L. Li et al., "Design of vertically-stacked polychromatic light-emitting diodes," Opt Express 17 (12), 9873-9878 (2009).

[8] W. Y. Fu, K. N. Hui, X. H. Wang et al., "Geometrical Shaping of InGaN Light-Emitting Diodes by Laser Micromachining," IEEE Photonics Technol Lett 21 (15), 1078-1080 (2009).

[9] K. N. Hui, K. S. Hui, H. Lee et al., "Enhanced light output of angled sidewall light-emitting diodes with reflective silver films," Thin Solid Films 519 (8), 2504-2507 (2011).

[10] K. N. Hui and K. S. Hui, "Vertically-stacked LEDs with invariance of color Chromaticity," Curr Appl Phys 11 (3), 662-666 (2011).

[11] X. H. Wang, P. T. Lai, and H. W. Choi, "Laser micromachining of optical microstructures with inclined sidewall profile," J Vac Sci Technol B 27 (3), 1048-1052 (2009).

[12] G. Y. Mak, E. Y. Lam, and H. W. Choi, "Precision laser micromachining of trenches in GaN on sapphire," J Vac Sci Technol B 28 (2), 380-385 (2010).

[13] K. J. Vampola, M. Iza, S. Keller et al., "Measurement of electron overflow in 450 nm InGaN light-emitting diode structures," Appl Phys Lett 94, 061116 (2009).

[14] J. J. Wierer, D. A. Steigerwald, M. R. Krames et al., "High-power AlGaInN flip-chip light-emitting diodes," Appl Phys Lett 78, 3379-3381 (2001).

[15] S. J. Park, M. K. Kwon, J. Y. Kim et al., "Enhanced emission efficiency of GaN/InGaN multiple quantum well light-emitting diode with an embedded photonic crystal," Appl Phys Lett 92, 251110 (2008).

[16] C. H. Kuo, C. M. Chen, C. W. Kuo et al., "Improvement of near-ultraviolet nitride-based light emitting diodes with mesh indium tin oxide contact layers," Appl Phys Lett 89, 201104 (2006).

[17] C. C. Kao, H. C. Kuo, H. W. Huang et al., "Light-output enhancement in a nitride-based light-emitting diode with 22 degrees undercut sidewalls," IEEE Photonics Technol Lett 17 (1), 19-21 (2005).

[18] F. K. Yam and Z. Hassan, "Innovative advances in LED technology," Microelectron J 36 (2), 129-137 (2005).

[19] L. Zhu, X. H. Wang, P. T. Lai et al., "Angularly Uniform White Light-Emitting Diodes Using an Integrated Reflector Cup," I IEEE Photonics Technol Lett 22 (7), 513-515 (2010).

[20] C. Hoelen, J. Ansems, P. Deurenberg et al., "Color tunable LED spot lighting, " SPIE 6337, 63370 (2006).

[21] J. K. Sheu, S. J. Chang, C. H. Kuo et al., "White-light emission from near UV InGaN-GaN LED chip precoated with blue/green/red phosphors," IEEE Photonics Technol. Lett. 15 (1), 18-20 (2003).

[22] G. Heliotis, P. N. Stavrinou, D. D. C. Bradley et al., "Spectral conversion of InGaN ultraviolet microarray light-emitting diodes using fluorene-based red-, green-, blue-, and white-light-emitting polymer overlayer films," Appl Phys Lett 87, 103505 (2005).

[23] Y. Xuan, D. C. Pan, N. Zhao et al., "White electroluminescence from a poly(N-vinyl-carbazole) layer doped with CdSe/CdS core-shell quantum dots," Nanotechnology 17, 4966-4969 (2006).

[24] Narendran, N, Maliyagoda, N, & Deng, L. et al., "Characterizing LEDs for general illumination applications: Mixed-color and phosphor-based white sources," SPIE 4445, 137-147 (2001).

Chapter 8

Nanolithography

Gunasekaran Venugopal and Sang-Jae Kim

Additional information is available at the end of the chapter

1. Introduction

Nanolithography is the branch of nanotechnology concerned with the study and application of the nanofabrication of nanometer-scale structures, meaning nanopatterning with at least one lateral dimension between the size of an individual atom and approximately 100 nm. The term nanolithography is derived from the Greek words "nanos", meaning dwarf; "lithos", meaning rock or stone; and "graphein" meaning to write. Therefore the literal translation is "tiny writing on stone", however nowadays one understands something different whenever this term is associated with nanotechnology. Nanolithography is used e.g. during the nanofabrication of leading-edge semiconductor integrated circuits (nanocircuitry), for nanoelectromechanical systems (NEMS) or for almost any other fundamental application across various scientific disciplines in nanoresearch.

This technology can be suitable to use in nanofabrication of various semiconducting Integrated Circuits (ICs), NEMS and for various applications in research. The modification in semiconductor chips at the nano-scale (in the range of 10^{-9} meter) is also possible. This method is contrasting to various existing nanolithographic techniques like Photolithography (Venugopal, 2011), Nanoimprint lithography (NIL), Scanning Probe Lithography (SPL), Atomic Force Microscope (AFM) nanolithography, Extreme Ultraviolet Lithography (EUVL) and X-ray Lithography.

In this chapter, the various nanolithographic fabrication techniques will be discussed in detail in which we will focus the various nano-patterning techniques/procedures suitable for device fabrication and their engineering applications. This technique is mainly used for nanofabrication. Nanofabrication is the method in which the devices can be designed and manufactured with the dimensions in nanometers [Kim, 1999; Venugopal, 2011a, 2011b, 2011c].

The conventional fabrication techniques like Focused Ion Beam (FIB) and wet etching methods are able to remove or etch the parts in the range to micron scale (Kim, 2001). However, in recent days, patterning and etching have to be done in nanoscale for specific applications. For that nano-fabrication and nano-level manipulation are the options to choose. Nanomanupulation plays major role in the field of nanofabrication. Nanomanupulation is a technique in which some specific tools are used to manipulate the objects in nanoscale (Parikh, 2008). At present, Scanning Probe microscopic methods involved in AFM [Davis, 2003) and Scanning Tunneling Microscopy (STM) are being used to manipulate the objects in nanometer scale. Specifically, AFM is being used to move the atoms, carbon nanotubes, nanoparticles, various nano-scale objects and also to test integrated circuits. Instruments used in nanolithography include the Scanning Probe Microscope (SPM) and the AFM. The SPM allows surface viewing in fine detail without necessarily modifying it. Either the SPM or the AFM can be used to etch, write, or print on a surface in single-atom dimensions (Venugopal, 2012).

The main drawbacks in the existing lithographic techniques will be carefully analyzed in this chapter. Also the need of nano-patterning for the low-cost, high throughput surface patterning technologies will be presented in this chapter. In addition, the complete coverage of nanolithographic process which includes Introduction, Resists and Masks, Photon-based Lithography, Electron Beam Lithography, Ion Beam Lithography and emerging nanolithographic techniques will be discussed in detail. However, the alternate nanolithography techniques like Micro-contact printing, Nanoimprint Lithography, Scanned Probe Lithography, Dip-pen Lithography will also be discussed in detail in this chapter

2. Importance of micro/nano patterning

Micro/Nano patterning is a one of the miniaturization technique for patterns, especially used for electronics. Nowadays it becomes a standard in biomaterials engineering and for fundamental research on cellular biology by mean of soft lithography. It generally uses photolithography methods but many techniques have been developed. The batch fabrication of microstructures requires a low-cost, high throughput surface patterning technology.

For example, it is important to design nanodevices such as nano-transistors and nanodiodes, nanoswitches and nanologic gates, in order to design nanoscale computers with tera-scale capabilities. All living biological systems function due to molecular interactions of different subsystems. The molecular building blocks (proteins and nucleic acids, lipids and carbohydrates, DNA and RNA) can be viewed as inspiring possible strategy on how to design high-performance NEMS and MEMS that possess the properties and characteristics needed. In addition, analytical and numerical methods are available to analyze the dynamics and three-dimensional geometry, bonding, and other features of atoms and molecules. So, electromagnetic and mechanical, as well as other physical and chemical properties can be studied. Nanostructures and nanosystems can be widely used in medicine and health. Among possible applications of nanotechnology are: drug synthesis and drug delivery (the therapeutic potential will be enormously enhanced due to direct effective delivery of new types of drugs to the specified body sites), nanosurgery and nanotherapy, genome synthesis and diagnostics, nanoscale actuators and sensors (disease diagnosis and prevention), nonrejectable nanoartificial organs design and implant, and design of high-performance nanomaterials.

It is important that these technologies drastically change the fabrication and manufacturing of materials, devices, and systems via:

- higher degree of safety

- environmental competitiveness.

- improved stability and robustness

- higher degree of efficiency and capability, flexibility and integrity supportability and affordability, survivability and redundancy

- Predictable properties of nano composites and materials (e.g., light weight and high strength, thermal stability, low volume and size.

3. Classification of lithographic techniques

There are many techniques through which micro/nano patterning could be possible. They are,

- Photolithography – an conventional and classical method

  o Ion beam Lithography

- X-ray lithography

- Electron beam lithography

- Alternate Nanolithographic Techniques

  o Micro-contact printing

  o Nano-imprint lithography

  o Scanning Probe lithography

3.1. Photolithography – A conventional and classical method

Lithography consists of patterning substrate by employing the interaction of beams of photons or particles with materials. Photolithography is widely used in the integrated circuits (ICs) manufacturing. The process of IC manufacturing consists of a series of 10-20 steps or more, called mask layers where layers of materials coated with resists are patterned then transferred onto the material layer.

A photolithography system consists of a light source, a mask, and an optical projection system. Photoresists are radiation sensitive materials that usually consist of a photo-sensitive compound, a polymeric backbone, and a solvent. Resists can be classified upon their solubility after exposure into: positive resists (solubility of exposed area increases) and negative resists (solubility of exposed area decreases). Fig. 1 shows the schematic of lithographic process in order to make the pattern on the desired substrate.

3.1.1. Patterning graphene device using photolithography

As mechanically exfoliated graphene sheets are in a mesoscopic scale, a lithographic technique is required to make metallic contacts on the sheet. Highly oriented pyrolytic graphite (HOPG) was used as the source material for graphene fabrication. Graphene flakes were mechanically transferred onto a highly doped silicon wafer. The graphene flakes for device fabrication were chosen by color and contrast method. p-type silicon wafers (100) with a boron doping concentration of $N_A = 10^{15}$ cm^{-3} can be used in which SiO_2 was thermally oxidized with the thickness of $t_{ox} = 300$ nm. The substrate, p +Si (resistivity 1-30 Ω cm), serves as a back-gate for the FET. To keep the disorder level comparable, standard RCA cleaning process followed by acetone and isopropyl alcohol to clean the Si/SiO$_2$ wafers.

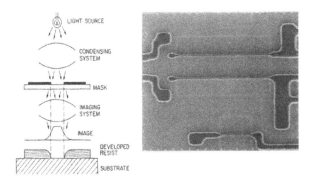

Figure 1. Schematic of photolithographic process. A pattern has been made on the substrate. (Scale bar is not mentioned)

Photo-lithography method can be used in this work to make electrode pattern. Details about the lithography process is discussed below. Figure 2 shows the mask aligner system.

Figure 2. Mask aligner system (MDA-400 M) for lithography pattern fabrication

The various stages of this lithography process or the procedures to be followed for lithographic pattern which are given below:

Stage -1: Wafer or substrate cleaning:

- Use clouse in the entire experiment
- Use Iso-propyl alcohol to clean wafer/substrate
- Use DI water to clean
- Use tissue paper and Air - drying to remove the water particles from the surface of wafer (both side).

Stage – 2: Spin coating of Photo-resist:

- The cleaned wafer to be put in spin coater and start creating vacuum
- Set spin coating rpm and time, using timer 1 and 2. (see the optimized parameter table)
- Put photo resist at the center of the cleaned wafer and spin coat.

Stage – 3: Baking the wafer

Put the spin coated wafer in the hot plate which is in 60º C for 150 sec and then remove wafer from the hot plate and do air cooling (only for back side of wafer)

Stage – 4: UV Exposure

Check and ensure the initial machine set up parameters is done carefully.

- Put the wafer on the substrate stage properly and press substrate Vac. button
- Put the mask in mask- holder and press Mask Vac. button on
- Mount the mask- holder over the substrate stage.
- Use Micrometer handle to bring the substrate stage and mask holder to touch each other. (no gap should be maintained between them). Be careful while doing this process.
- Press Vacuum contact button
- Then wait for 20 sec.
- Turn Align and Exposure knob one by one carefully.
- Now the aligner system will start working and comes to its position.
- Set the expose time by using timer (standard 2.5 sec)
- Now press Exposure button.
- After exposing UV light, turn off align and exposure knob.

Stage – 5: Removal of wafer

- Now put off Vacuum contact button

- Then bring down the stage by using Micrometer handle
- Remove mask holder carefully
- Then press substrate vac. button off and remove wafer from the wafer stage.

Stage – 6: Developing process

The UV-exposed wafer to be put in developer solvent [standard developer solution AZ 300 MIF used]. Slow soaking has to be performed with respect to user need and process. In this process, the UV-unexposed parts (in case of positive PR) the photo-resist will be dissolved in the developer solution and show clear electrode pattern fabricated via mask. Then put the wafer in DI water bath for 1 min and do air- drying to clean the wafer thoroughly.

Stage – 7: Pattern Analysis

After developing,

- Put the developed wafer again in wafer stage and press substrate Vac. Button
- Use CCD camera -module's adjusting knobs in order to check the developed pattern.

Sample	Spin coating		Prebake		Expose UV with Mask	Developing time	Remarks
1	300 rpm 6sec (timer 1)	2200 rpm 40 sec (timer 2)	60°C	150 sec	55 sec	2 sec	Excellent 4-probe Pattern obtained

Table 1. Optimized condition for Lithographic Process in MDA-400M Mask Aligner (using AZ 5214 positive photo-resist)

The schematic of the detailed lithographic process is presented in Figure.3 (a-e).

After completing the experiment: Mask cleaning to be done

Mask cleaning after exposure is important in the lithographic pattern process. Hence, the defects deposited in mask can be avoided during next experiment time. To clean the mask, the following things to be followed:

- rinse the mask with acetone
- rinse the Mask with IPA
- Dry the Mask using the N_2 gun

The Lithographic process was followed which was described in the above section 2.4. The positive photo-resist (AZ 5214) was spin-coated over the graphene flakes on the substrate. By using photolithography (Mask Aligner MDA- 400M; MIDAS), the graphene flakes were patterned through *Cr* mask for electrode formation. Then the gold (99.99 %) electrodes of 100 nm- thick were formed through thermal evaporation technique and structured by lift-off using acetone. A metal contact was made to the substrate as the back-gate contact. After lift-off

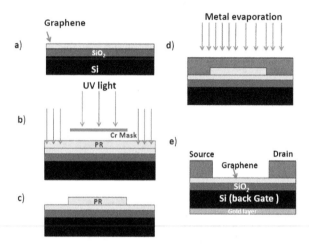

Figure 3. Photolithographic process of electrode patterning on graphene. (a) Graphene flake on Si/SiO₂ substrate (b) Photoresist is spin-coated over the graphene flake and UV light illuminated through Cr mask. (c) the pattern after developing process (d) gold (Au) evaporation through thermal evaporation technique (e) after lift-off process, the device with source and drain electrode structure with back-gate configuration. (Venugopal, 2011).

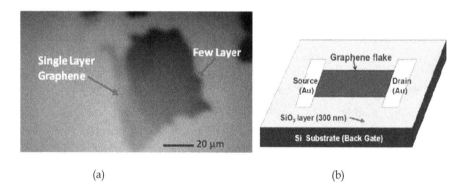

         (a)   (b)

Figure 4. (a) An optical image of single layer graphene is shown. Single and few layer graphene flakes are clearly seen. (b) Schematic of graphene device with electrode pattern.

process, the device was annealed at 200° C in Ar/H₂ atmosphere for 45 min to improve the adhesion with graphene flake as well as to avoid contaminants. After the lithography, metallic Au/Al electrodes are deposited by using thermal evaporation system. Then lift-off process is carried out (using acetone) to get the final pattern for device characterization. If necessary, graphene can be etched to a desired shape by the oxygen plasma ashing with negative or positive electron-beam resist stencils, which were not followed here.

3.2. Ion beam lithography

In this section, ion beam based patterning technique is discussed. For example Focused Ion Beam (FIB) based three dimensional etching method is followed for patterning micro/nano devices (Seliger, 1979).

3.2.1. Focused ion beam 3-D fabrication technique

Miniaturization is the central theme in modern fabrication technology. Many of the components used in modern products are becoming smaller and smaller. Here, the focused ion beam (FIB) direct milling technique will be discussed with the focus on fabricating devices at the micrometer to nano-scale level. Because of the very short wavelength and very large energy density, the FIB has the ability for direct fabrication of structures that have feature sizes at or below 1 μm. As a result, the FIB has recently become a popular candidate in making high-quality microdevices or high-precision microstructures (Kim, 2008).

The FIB has been a powerful tool in the semiconductor industry mainly for mask repairing, device modification, failure analysis and integrated circuit debugging. Two basic working modes, ion beam direct write and ion beam projection, have been developed for these applications. The ion beam direct write process, also known as FIB milling (FIBM), is the process of transferring patterns by direct impingement of the ion beam on the substrate. It is a large collection of microfabrication techniques that removes materials from a substrate and has been successfully used for fabricating various three-dimensional (3D) micro structures and devices from a wide range of materials. For the ion beam projection process, a collimated beam of ions passes through a stencil mask and the reduced image of the mask is projected onto the substrate underneath. The ion beam projection process is also known as focused ion beam lithography (FIBL) and can serve as an alternative to conventional optical lithography (Kim, 1999).

For example, to develop the graphite stacked-junctions, planar-type nanostructures, a high-resolution FIB instrument (SII SMI-2050) can be used. The photo-image of FIB unit and the schematic of FIB functions are presented in Figure. 5.

The 3-D etching technique can be followed by tilting the substrate stage up to 90° automatically for etching thin graphite flake. It has freedom to tilt the substrate stage up to 60° and rotate up to 360°. The steps of the fabrication process using a FIB etching are shown in Figure. 6 (a–d). The clear axes of the FIB process configurations with in-plane (x–y) and vertical axes (as z direction) are indicated in an axis diagram in Figure. 6(b). The in-plane area was defined by tilting the sample stage by 30° anticlockwise with respect to the ion beam and milling along the ab-plane. The in-plane etching process is shown in Figure. 6(a)-(c). The out of plane or the c-axis plane was fabricated by rotating the sample stage by an angle of 180°, then tilting by 60° anticlockwise with respect to the ion beam, and milling along the c-axis direction (Saini, 2010). The schematic diagram of the fabrication process for the side-plane is shown in Figure. 6(d).

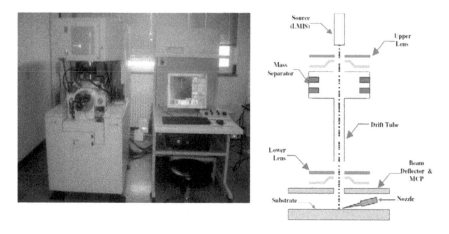

Figure 5. The photograph image of FIB unit and schematic of FIB machine

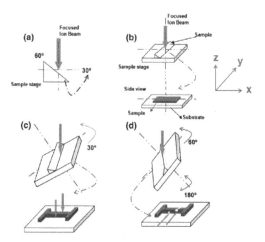

Figure 6. Nanoscale stack fabrication process using focused ion beam 3D etching method. (a) Scheme of the inclined plane has an angle of 60° with ion beam (where we mount sample). (b) The initial orientation of sample and sample stage. (c) Sample stage titled by 30° anticlockwise with respect to ion beam and milling along *ab*-plane. (d) The sample stage rotated by an angle of 180° and also tilted by 60° anticlockwise with respect to ion beam and milled along the c-axis. (Venugopal, 2011d)

By varying the stack height length and in-plane area, the various sizes of stacked-junctions can be fabricated on the graphite layer. The number of elementary junctions in the stack will vary depends on the height of the junction. If junction height is more, the more number of elementary junctions exists which provide larger more resistance in c-axis characteristics.

3.3. X- ray lithography

This lithography processes involve the category of nanolithographic techniques, through which transistors with smaller features can be patterned. It uses X-rays to transfer a geometric pattern from a mask to a light-sensitive chemical photoresist, or simply "resist," on the substrate. A series of chemical treatments then engraves the produced pattern into the material underneath the photoresist.

X-ray lithography can be extended to an optical resolution of 15 nm by using the short wavelengths of 1 nm for the illumination. This is implemented by the proximity printing approach. The technique is developed to the extent of batch processing. The extension of the method relies on Near Field X-rays in Fresnel diffraction: a clear mask feature is "demagnified" by proximity to a wafer that is set near to a "Critical Condition". This Condition determines the mask-to-wafer Gap and depends on both the size of the clear mask feature and on the wavelength. The method is simple because it requires no lenses. This technique originated as a candidate for next-generation lithography for the semiconductor industry, with batches of microprocessors successfully produced. Having short wavelengths (below 1 nm), X-rays overcome the diffraction limits of optical lithography, allowing smaller feature sizes. If the X-ray source isn't collimated, as with a synchrotron radiation, elementary collimating mirrors or diffractive lenses are used in the place of the refractive lenses used in optics. Fig. 7 illustrate the process of X-ray lithography.

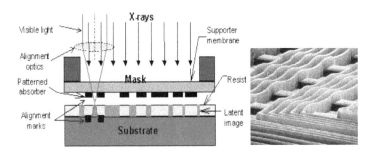

Figure 7. (a) Schematic process of X-ray lithography. (b) structure produced with X-ray lithography (Courtesy Source : SAL, Inc.). (Scale bar is not mentioned)

The X-rays illuminate a mask placed in proximity of a resist-coated wafer. The X-rays are broadband, typically from a compactsynchrotron radiation source, allowing rapid exposure. Deep X-ray lithography (DXRL) uses yet shorter wavelengths on the order of 0.1 nm and modified procedures such as the LIGA process, to fabricate deep and even three-dimensional structures. X-rays are usually generate secondary electrons as in the cases of extreme ultraviolet lithography and electron beam lithography. While the fine pattern definition is due principally to secondaries from Auger electrons with a short path length, the primary electrons will sensitize the resist over a larger region than the X-ray exposure.

3.3.1. Advantages of X-ray lithography

There are several advantages in X-ray lithography.

1. Resolves diffraction issues

2. Shorter wavelengths (0.1 - 10 nm) can be used

3. Smaller features can be patterned

3.3.2. Disadvantages of X-ray Lithography

The following are the disadvantages of X-ray Lithography

1. Usage of X-ray masks

2. Deformation during the process

3. Vibrations during the process

4. Time consuming process

3.4. E-beam lithography

Electron Beam Lithography uses a tightly focussed beam of electrons scanned over the surface of a substrate. Typically, electron beam lithography with ultra high resolution (UHR) is used at the very beginning of a multiple technique and a multiple step process in a top down approach in order to transfer the nanostructure into the substrate or subsequently build up a device in a layer by layer fashion.

3.4.1. E-beam applications

This E-beam lithographic technique is mainly having following advantages in research field:

1. Research and Development

2. Advanced processing techniques

3. Future processing equipment

4. Can convert SEM to be used as an EBL machine

5. Minimum resolution is slightly larger

6. Used with photolithography and X-ray lithography to create next generation devices.

For nanolithography with ultra high resolution down to sub10nm feature sizes, complete dedicated e-beam writer systems or converted scanning electron microscopes (SEM) can be used. With the help of a design editor and a pattern generator, the electron beam is guided over the substrate surface, which is covered with electron beam sensitive resist such as PMMA, in order to generate a resist mask which then can be further used for nanopattern transfer. The steps of e-beam lithography is given in Fig. 8.

a. Resist Preparation

In this Process, the PMMA solution is spin coated onto the sample and baked to harden the film and remove any remaining solvent.

b. Exposure

Selected areas of sample are exposed to a beam of high energy electrons

c. Development

Sample is immersed in developer solution to selectively remove resist from the exposed area.

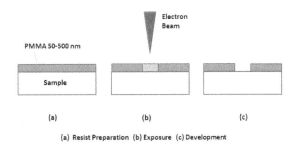

Figure 8. Schematic process of e-beam lithography. (a) Resist Preparation (b) Exposure (c) Development

3.4.2. E-beam lithography advantages

1. The resolution is not limited by diffraction; minimum feature is written on the nanoscale.

2. Can write smaller features than X-ray lithography and photolithography

3. Pattern is written directly to the wafer.

4. Used to develop specialized devices and prototype devices

5. Fast turn-around time

6. This employs a beam of electron instead of photons

3.4.3. E-Beam Lithography Disadvantages

1. Not an efficient process for industrial processing

2. Takes multiple hours to pattern entire wafer

3. Machines are costly

4. Greater than 5 million dollars

5. System is more complex than photolithography system

6. Scattering and over exposure result in minimum feature being larger

7. Slow throughput

3.5. Alternate nanolithographic techniques

3.5.1. Micro-contact printing (soft lithography)

This is known as soft lithography that usually uses the relief patterns on a PDMS (poly-dimethylsiloxane) stamp in order to form patterns of self-assembled monolayers (SAMs) of ink on the surface of a substrate through conformal contact. This technique has wide range of application in cell biology, microelectronics, surface chemistry, micromachining, Patterning cells, patterning DNA and Patterning protein.

Figure 9 represents the process of Micro-contact printing. This process involves the application of ink to stamp, application of stamp to surface, removal of stamp and residues rinsed off

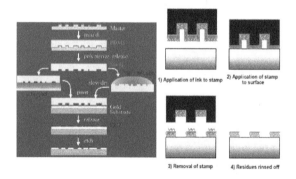

Figure 9. Schematic of the Micro-contact printing process (Courtesy: IBM Zurich)

Advantages of Micro-contact printing:

1. Very simple and easy pattern procedures to create micro-scale features

2. This can be done in a traditional laboratory environment. No need clean room facility

3. Using single master, multiple stamps can be made

4. Reliability of individual stamps which can be used for many times.

5. It is a cheaper method.

Limitation in Micro-contact Printing:

1. Diffusion of ink from PDMS stamp to surface during pattering

2. Shrinking of stamp is one of main problem in which stamp can eventually shrink in size resulting difference in desired dimensions of the substrate patterning

3.   Contamination of substrate

4.   Stamp deformation

3.5.2. Nano-imprint lithography

Nanoimprint lithography (NIL) is an emerging process that can produce sub-10nm features. It is a simple process that uses a mould to emboss the resist with the required pattern. After embossing the resist, compressed resist material is removed using anisotropic etching and the substrate exposed. It can produce features at extremely small resolutions that cover a large area with a high throughput and relatively low cost, which is main advantage of this technique. It can be adapted to transfer all components needed to create a thin film transistor on a plastic substrate. It involves pressing and heating a thin film between a patterned template and a substrate. Upon heating, the patterned film adheres only to the substrate [ref Fig. 10]. This has high throughput and is relatively inexpensive compared to developing extreme deep UV lithography for commercial viability. It is also flexible enough to be used at chip level with several layers or at the wafer level when single layer is required. It can give resolutions lower than 10nm with high throughput at low cost. One of the current barriers to production at these resolutions is the development of mould. It can be used for fabricating nanoscale photo-detectors, silicon quantum-dot, quantum wire and ring transistors (Chou, S.Y. 1996)

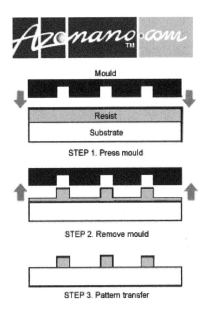

Figure 10. Schematic diagram of the steps involved in the nanoimprint lithographic process [Courtesy: Source: AZo-Nano]

Applications of Nanoimprint lithography

1. It can be used to make optical, photonic, electrical and biological devices.

2. Advances in mould manufacturing will have wide application of NIL in smaller devices.

3.5.3. Scanning Probe Lithography (SPL)

SPL is an emerging area of research in which the scanning tunneling microscope (STM) or the atomic force microscope (AFM) is used to pattern nanometer-scale features. The patterning methods include mechanical pattering such as scratching or nano-indentation, or local heating with sharp tip (Dagata, 1995). When a voltage bias is applied between a sharp probe tip and a sample, an intense electric field is generated in the vicinity of the tip (Ref. Fig. 11).

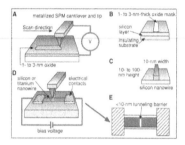

Figure 11. Schematic diagram of the Scanning probe lithography. Electrical bias between a conducting tip and a substrate induces a highly localized enhanced oxidation [Courtesy Source: Dagata et al, *Science*, Vol.270, pp.1625-1626, 1995]

Advantages:

1. This process can make nanopatterns without optical apparatus.

2. It can control deposited material by hydrophobicity of the surface.

3. This process can make arbitrary patterns by controlling the trajectory of AFM tip.

4. This process involves small scan area, Low throughout.

5. By using this technique, it is possible to make nanowire, SET, etc.. By using an organic solvent, the organic material can be deposited on the surface.

SPL method is a recognized as a lithographic tool in the deep sub-micron regime, as it is compatible with standard semiconductor processing. There are four main factors which dictate the viability of SPL as a known patterning method for the semiconductor industry. They are

1. Throughput (wafers/hour)

2. Resolution

3. Alignment accuracy

4.   Reliability

Scanning probe lithography involves a set of lithographic techniques, in which a microscopic or nanoscopic stylus is moved mechanically across a surface to form a pattern. In this method, another technique describes a SPL technique which is known as Dip Pen Nanolithography.

Dip Pen Nanolithography - in this process, the patterning is done by directly transferring chemical species to the surface. We can call this process as constructive process.

3.5.3.1. Dip Pen Nanolithography (DPN)

Dip Pen Nanolithography (DPN) is known as a soft-lithography technique that uses an AFM scanning probe tip to draw nanostructures. In this process, a probe tip is coated with liquid ink, which then flows onto the surface to make patterns wherever the tip makes contact. This kind of directwrite technique provides high-resolution patterning capabilities for a number of molecular and biomolecular "inks" on a variety of substrates. Substrates are the base material that the images are printed on. Some of the applications of the DPN technique include sol gel templates that are used to prepare nanotubes and nanowires, and protein nanoarrays to detect the amount of proteins in biological samples such as blood. (Ref. Fig. 12).

This process was first developed by Professor Chad Mirkin at Northwestern University Nanotechnology Institute for depositing thin organic films in patterns with feature sizes of around 10 nm (about 20 times better than the best optical lithography) (Mirkin, 1999).

In DPN technology, the ink on a sharp object is transported to a paper substrate via capillary forces. The capillary transport of molecules from the AFM tip to the solid substrate is used in DPN to directly "write" pattern consisting of a relatively small collection of molecules in nanometer dimesions. An AFM tip is used to write alkanethiols with 30-nm line width resolution on a gold thin film in a manner analogous to that of a dip pen. Molecules are delivered from the AFM tip to a solid substrate of interest via capillary transport, making DPN a potentially useful tool for creating and functionalizing nanoscale devices (Mirkin, 1999).

Several factors decide the resolution of DPN:

1.   The grain size of the substrate affects DPN resolution much as the texture of paper controls the resolution of conventional writing.

2.   The tip-substrate contact time and thus the scan speed influence DPN resolution.

3.   Chemisorption and self-assembly of the molecules can be used to limit the diffusion of the molecules after deposition.

4.   Relative humidity seems to affect the resolution of the lithographic process by controlling the rate of ODT transport from the tip to the substrate. The size of the water meniscus that bridges the tip and substrate depends on relative humidity. For example, the 30-nm wide line required 5 min to generate in a 34% relative humidity environment, whereas the 100-nm line required 1.5 min to generate in a 42% relative humidity environment.

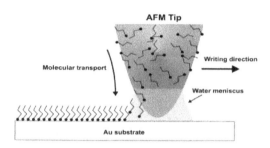

Figure 12. Schematic representation of DPN. A water meniscus forms between the AFM tip coated with ODT and the Au substrate. The size of the meniscus, which is controlled by relative humidity, affects the ODT transport rate, the effective tip-substrate contact area and DPN resolution. [Courtesy: C.A. Mirkin, et al, Science 283, 661, 1999]

3.5.3.2. DPN application on semiconductor surfaces

Dip-Pen Nanolithography can not only apply to gold surface using alkyl or aryl thiols as inks, but also to semiconductor surfaces, such as silicon and gallium arsenide. Hexamethyldisilazane (HMDS) is used as the ink to pattern and modify (polarity) the surface of semiconductors. Lateral force microscopy (LFM) can be used to differentiate between oxidized semiconductor surfaces and patterned areas with the deposited monolayers of HMDS. The choice of the silazane ink is a critical component of the process since the traditional adsorbates such as trichlorosilanes are incompatible with the water meniscus and polymerize during ink deposition. This work provides insight into additional factors, such as temperature and adsorbate reactivity, that control the rate of the DPN process and paves the way for researchers to interface organic and biological structures generated via DPN with electronically important semiconductor substrates (Ivanisevic, 2001).

3.5.3.3. DPN application on magnetic materials: Approach to high density recording and storage

Over the past decade, there has been considerable interest in methods for synthesizing and patterning nanoscale magnetic materials. These nanomaterials show novel size-dependent properties, are potentially useful for high-density recording. Two of the main challenges in this field are: (a) site-and shape-specific patterning of hard magnetic nanostructure on the sub-100 nm scale; (b) ability to reliably and reproducibly read/write such minute features. The conventional top-down approach in recording media is plagued by the difficulties of etching and patterning novel hard magnetic systems, especially as the individual recording elements approach the super paramagnetic limit at room temperature operations. DPN can be used as a direct-write method for fabricating "hard" magnetic barium hexaferrite, $BaFe_{12}O_{19}$ (BaFe), nanostructures. This method utilizes a conventional atomic force microscope tip, coated with the BaFe precursor solution, to generate patterns that can be post-treated at elevated temperature to generate magnetic features consisting of barium ferrite in its hexagonal magnetoplumbite (M-type) structure. Features ranging from several hundred nm down to below 100 nm can be generated. (Fu, 2003).

4. Conclusion

In conclusion, the complete nanolithographic processes which include introduction, resists and masks, Photon-based lithography, electron beam lithography, ion beam lithography and emerging nanolithographic techniques like the alternate nanolithography techniques Micro-contact printing, Nanoimprint Lithography, Scanned Probe Lithography, Dip-pen Lithography were briefly discussed in this chapter. As a conclusion, the following table can presents the complete scenerio of the nanolithographic process.

	Ideal	Advanced Optical	Electron Beam	Nano Imprint
Pattern shape	Any	Any	any	Any
Resolution (nm)	Good	Good	Good	Good
Alignment (nm)	Good	Good	Good	Poor
Large & small Pattern	Yes	Yes	No	Yes
Throughput (wafers/hr)	High	High	Low	Medium
Running Cost	Low	High	Low	Medium
Initial Cost	Low	High	Medium	Low

Table 2. Comparison of results between conventional and Nanolithographic methods

Acknowledgements

The authors would like to acknowledge the supports from the National Research Foundation of Korea Grant under contract number 2011-0015829 and 2011 Jeju Sea Grant College Program funded by the Ministry of Land, Transport and Maritime Affairs (MLTM), Republic of Korea. A part of this work was carried out at the Research Instrument Center (RIC) at Jeju National University, Jeju, and Republic of Korea. Also the author (G.V) would like to thank the management of Karunya University for continuing this research through their KSTG seed money grant No. KU/Reg/D(R)/OO/12/29 dated 21.01.2012 and DST NanoMission, New Delhi.

Author details

Gunasekaran Venugopal[1] and Sang-Jae Kim[2]

1 Karunya University, Department of Nanosciences and Technology, Tamil Nadu, India

2 Jeju National University, Department of Mechatronics Engineering, Jeju, South Korea

References

[1] Chou, S. Y. (1996). Nano imprint Lithography. *J. Vac. Sci. Technol. B.,* , 14, 4129.

[2] Dagata, J A. (1995). Device Fabrication by Scanned Probe Oxidation. *Science,* , 270, 1625-1626.

[3] Davis, R. C. (2003). Chemomechanical surface patterning and functionalization of silicon surfaces using an atomic force microscope. Appl. Phys. Lett. , 82

[4] Fu, L. (2003). Dip-Pen Patterning of magnetic materials. Nano Lett., , 3, 757-760.

[5] Ivanisevic, A. (2001). Dip-pen nanolithography on semiconductor surfaces. J. Am. Chem. Soc. , 123, 7887-7889.

[6] Kim, S. J. (1999). D intrinsic Josephson junctions using c-axis thin films and single crystals. *Supercond. Sci. Technol.,* , 12, 729-731.

[7] Kim, S. J. (1999). Submicron Stacked-junction Fabrication from $Bi_2Sr_2CaCu_2O_{8+\delta}$ Whiskers by Focused-Ion-Beam Etching. *Appl. Phys. Lett.,* , 74(8), 1156-1158.

[8] Kim, S. J. (2001). Fabrication and Characteristics of Submicron Tunnelling Junctions on High Tc Superconducting c-axis Thin Films and Single Crystals. *J. Appl. Phys.,* , 89(11), 7675-7677.

[9] Kim, S. J. (2008). Development of Focused Ion Beam Machining Systems for Fabricating Three-dimensional Structures. *Jpn. J. Appl. Phys.,* , 47(6), 5120-5122.

[10] Mirkin, C. A. (1999). Dip-pen Nanolithography. Science. , 283

[11] Parikh, D. (2008). Nanoscale Pattern Definition on Nonplanar Surfaces Using Ion Beam Proximity Lithography and Conformal Plasma-Deposited Resist. J. Microelectromech. Syst. , 17

[12] Saini, S. (2010). Characterization of Submicron Sized Josephson Junction Fabricated in a $Bi_2Sr_2Ca_2Cu_3O_{10+\delta}$ (Bi-2223) Single Crystal Whisker. *J. Supercond. Nov. Magn.* , 23, 811-813.

[13] Seliger, R. L. (1979). High-resolution, Ion-beam Processes for Microstructure Fabrication. *J. Vac. Sci. Technol. B.,* , 16(6), 1610-1612.

[14] Venugopal, G. (2011). Temperature dependent transfer characteristics of graphene field effect transistors fabricated using photolithography. *Curr. Appl. Phys.* , 11, S381-S384.

[15] Venugopal, G. (2011a). Fabrication of Nanoscale Three-dimensional Graphite Stacked-junctions by Focused-ion-beam and Observation of Anomalous Transport Characteristics. *Carbon,* , 49(8), 2766-2772.

[16] Venugopal, G. (2011b). Temperature Dependence of Transport Anisotropy of Planar-type Graphite Nano-structures Fabricated by Focused Ion Beam. *J. Nanosci. Nanotechnol.* , 11(1), 296-300.

[17] Venugopal, G. (2011c). Fabrication and Characteristics of Submicron Stacked-Junctions on Thin Graphite Flakes. *J.Nanosci. Nanotechnol.* , 11(2), 1405-1408.

[18] Venugopal, G. (2011d). Investigation of electrical transport characteristics of nano-scale stacks fabricated on thin graphite layer. *Thin Solid Films.* , 519, 7095-7099.

[19] Venugopal, G. (2012). An investigation of the electrical transport properties of graphene-oxide thin Films. *Matl. Chem. Phys.* , 132, 29-33.

Chapter 9

Nanotechnology for Packaging

Artin Petrossians, John J. Whalen III,
James D. Weiland and Florian Mansfeld

Additional information is available at the end of the chapter

1. Introduction

The use of implantable microelectronic devices for treatment of medical conditions, e.g. movement disorders, deafness and urinary incontinence has increased steadily over the years [1]. These devices use microelectronic components to sense biological activities in the implanted patient. The microelectronic components must be protected from the surrounding tissue using insulating (hermetic) packaging material. This packaging prevents the aqueous saline environment of the body from corroding, short-circuiting and contaminating the internal electronics. Microelectronic packages must incorporate some electrically conducting elements that bridge through the protective packaging to allow the internal microelectronics to sense (or stimulate) the surrounding external environment. These conductive elements are called interconnects or feed-throughs.

First generation implantable electronic devices, such as the first cardiac pacemakers, had few, relatively large interconnects. To this day, many of these interconnects are constructed using labor intensive, manual assembly techniques. Electrode contacts are tack-welded to conductive leads, and then the entire assembly is laid inside a mold and encased in silicone insulation. The insulated wires are then released from the mold, flashing is removed and the final assembly is quality tested.

Next generation devices could have as many as 1000 interconnects condensed in a similar cross-sectional surface area of the device packages. At this scale, hand laid and molded wires in silicone will not suffice. Simple but accurately repeatable processes must be developed to create functional feed-throughs. Existing feed-through designs and fabrication processes will not be appropriate and will warrant new strategies to prevent the penetration of mobile ions such as K^+, Na^+ and Cl^- from surrounding body fluid [2, 3].

The hermeticity of a package is its ability to prevent ion migration across (or through) its structure. Research and development in this area has accelerated significantly [4]. Simply put the hermeticity of different materials classes can be ordered as follows: polymers (least hermetic) < glass < polycrystalline metals/oxides (most hermetic). Interfaces between two different materials can serve as an avenue for contaminant ions to migrate as can grain boundaries between crystals of the same, polycrystalline material. Helium leak testing is currently the gold standard for evaluating hermeticity.

A number of research groups are trying to overcome existing issues with fabrication of hermetic packages for implantable microelectronics [5-8]. Since these implanted devices must remain hermetic for the lifetime of the patients, many important factors should be considered in their design, including the location of implantation, dimensional constraints, materials constraints (e.g. biocompatibility, conductivity, etc.), and selection of an appropriate fabrication process technology

Our group has investigated using electrochemical plating to fill cylindrical channels in channeled substrates. Our theoretical approach is to take a two-dimensional substrate with penetrating channels orthogonal to the planar surfaces, and fill these channels with electrochemically deposited metal. The resulting "assembly" of conducting elements embedded in the insulating substrate, can then be bump-bonded to a microchip, and the chip can then be encased in a gas-filled, brazed hard casket or embedded in a conformal coating.

This approach has been successful using larger channeled (Υ = 200 µm) substrates, like the U.S. Naval Research Labs channel glass, with electroplated copper and chromium interconnects [9]. However, electrochemically depositing non-porous, continuous interconnects made from implantable electrical stimulator metals (e.g. platinum, iridium or their alloys) cannot be achieved due to solubility and deposition rate challenges. As a result, our group proposed and developed a strategy of electroplating several adjacent high-aspect ratio nano-channels, in nano-channeled substrates, with precious metal solutions. Once deposited, adjacent embedded nanowires can be electrically connected in parallel to create larger conducting elements.

Figure 1 depicts two schematics illustrating how this type of assembly would work. Here, metallic nanowires are deposited into a nano-channeled substrate, forming hermetic feedthroughs. Instead of using a single conducting element bump-bonded to each single contact on the chip, an array of co-deposited nanowires forms a single conductor unit through the substrate.

Fabricating these assemblies is achieved using a commonly used approach by nanowire researchers, called the "template synthesis" approach [10]. In this method, metallic ions in solution (plating solution) are electrochemically reduced at a working electrode surface that has been applied to the base of the channels of a nano-channeled substrate.

Using this template synthesis approach to deposit nanoscale elements provides distinct performance advantages. By confining the metal deposition to nanometer dimensions inside the channel, mass transport gradients that cause dendritic deposition and growth no longer occur. Further, by completely filling the channels, we are able to develop a dense interconnect from platinum and iridium that spans the substrate material.

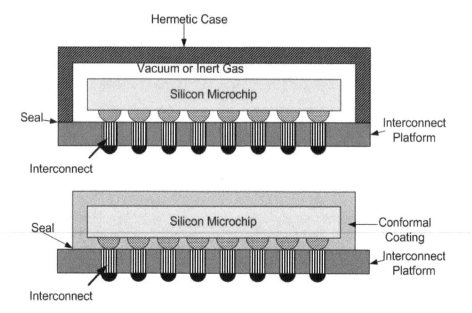

Figure 1. The top diagram shows an interconnect substrate with a case bonded over the top of the chip. Bumps on both the chip and the interconnect substrate facilitate electrical connections. This interconnect substrate could also be used with conformal coating technology (bottom). Drawing not to scale.

This approach also provides several key commercialization advantages. First, electroplating is a cost-effective processing technique. Second, unlike many nano-fabrication processes, template synthesis can be performed at ambient temperature and pressure. Lastly non-equilibrium phases can be produced by electrodeposition, a result that cannot be achieved using thermal processing techniques [11].

To date, no study has reported fabrication of ultra-high-density platinum-iridium nanowire arrays using a template synthesis approach. Only a handful of reports have been published on platinum nanowires synthesis. Approaches reported include focused ion beam [12], photoreduction in mesoporous silicides [13-17], colloidal synthesis [18, 19], self-assembly [20, 21], and nano-channel filling by electrochemistry [22-24],

This study focuses on the fabrication and evaluation of nonporous, platinum-iridium dense nanowires with improved electrical and mechanical properties to be used, embedded in their original template substrate, as a novel feed-through technology in hermetically packaged implantable microelectronics. Here we report on the fabrication process, the material properties of the isolated nanowires, and lastly an assessment of the performance characteristics of the nanowire-in-template assembly as a hermetic feed-through platform.

2. Materials and methods

2.1. Platinum–Iridium nanowire electrodeposition

Figure 2 is a schematic detailing the key steps involved in the template synthesis approach used to fabricate our platinum-iridium nanowires and nanowire-template assemblies. Prior to deposition, a conductive thin film layer must be applied to one side of the filtration membrane to 1) seal the base of the pores, thus allowing them to be filled with plating solution, and 2) to provide an electrically conducting base to serve as the working electrode at which metallic ions in solution are reduced and "grown" through the template as a metallic nanowire.

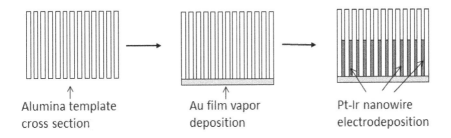

Alumina template   Au film vapor   Pt-Ir nanowire
cross section   deposition   electrodeposition

Figure 2. Schematic of the fabrication processes of metallic nanowires in AAO nanopores. Drawing not to scale.

Anodisc® nanoporous anodized aluminum filtration membranes (Whatman Inc., UK) were used as substrates for nanowire template synthesis. The templates have an approximate thickness of 60 μm and a diameter of 47 mm. For ease of membrane handling all templates were fitted with a plastic annular ring attached to one side of each membrane.

SEM inspection of both sides of the membranes revealed that the pore apertures sizes were different on one side versus the other. The pore apertures on the side with the plastic ring attached were approximately 20 nm in diameter. On the other side the pore apertures were approximately 200 nm in diameter, randomly distributed and with larger spaces between the pores. Cross-sectional analysis (through the membrane thickness) revealed that the large 200 nm channels continued down the majority of the template thickness, and that in the final 50 nm of thickness, the channels bifurcated into series of smaller finger-like channels with 20 nm diameters. Au thin films (h = 80 nm) were e-beam vapor-deposited on the side of the AAO membranes with the smaller pore apertures to create the sealed, electrical contact base for the working electrode.

Nanowires were electrodeposited using a three-electrode electrochemical cell which contained a larger, vertically-oriented cylindrical channel and a smaller diameter cylindrical chamber were machined into a Teflon® block along with horizontal small via to create a Luggin capillary between the two Figure 3. The larger channel's base was sealed by placing (in the following order,): an O-ring against the Teflon® block, followed by the AAO membrane with the uncoated side in contact with the O-ring, followed by a thick (h = 40 mm) copper plate. These components were fixated in place using a spring clamp. Once assembled, the electrochemical plating solution was filled into the larger chamber and the reference electrode was inserted into the smaller chamber.

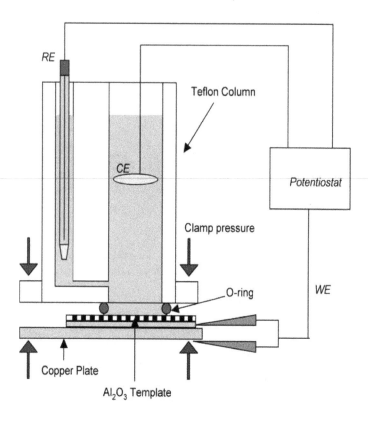

Figure 3. Electrochemical cell used for nanowire deposition in nano-channeled aluminum oxide (Al2O3) template.

Electrochemical deposition was performed using a software controlled, programmable potentiostat (Gamry). Electrical contact was made to the base copper plate, via alligator connector, thus making the sputtered Au thin film serve as the working electrode (WE). A silver-silver chloride (Ag/AgCl) electrode was used as the reference electrode (RE) and a spiral wound platinum wire ($\curlywedge$ = 1 mm) was placed in the larger chamber as a counter electrode (CE).

Nanowires were electrochemically deposited from a platinum-iridium plating solution that our group has developed and reported elsewhere [25]. In this work, two key parameters were controlled to affect deposition properties: pH and deposition potential. The pH of the initial plating solution is approximately 1.8 to 2.5, depending on the desired final properties of the deposited alloy, and in this study, was varied from pH = 1.8 to 5.0 by carefully titrating the solution with 3M NaOH (aq) solution. With respect to deposition potential, a potentiodynamic program was used to drive deposition. Previous studies by our group have shown that changing the potential range impacts the compositional ratio of Pt:Ir [25]. Specifically the potential range was cycled over a 150mV potential range, e.g. between U = 0.0 V to -0.15 V vs. Ag/AgCl. The ranges used are listed in Table 1.

2.2. Nanowire isolation

Nanowires were isolated from the AAO templates, Figure 4, for further analysis of the nanowire properties. The electrodeposited AAO templates with embedded nanowires were immersed in an aqueous solution of 3M NaOH(aq) to dissolve the oxide membrane. The nanowire suspension was allowed to stand, to settle the nanowires out of the basic solution. Excess solution (supernatant) was carefully pipetted off and DI water was added to the vial to neutralize the remaining supernatant's pH. This process was repeated three times until a neutral solution was achieved. Nanowires in suspension were then pipetted onto either fresh, un-sputtered AAO filters to capture for SEM or onto TEM mesh grids for TEM analysis.

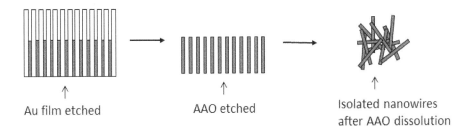

Au film etched   AAO etched   Isolated nanowires
                                                                after AAO dissolution

Figure 4. Schematic of the isolation process used to separate electrodeposited nanowires from AAO templates. Drawing not to scale.

2.3. Electron microscopy (SEM & TEM) characterization

All scanning electron microscopy imaging was performed using a field emission scanning electron microscope (ZEISS 1550VP) with an accelerating voltage of 4 kV.

Tranmission electron microscopy was used to further characterize nanowire morphology and microstructure. Isolated nanowires were captured on carbon coated copper grids with 300-mesh size (Ted Pella Inc.) and imaged using a JEOL 2100 (Japan) TEM. Brightfield and darkfield images as well as diffraction patterns were captured. Images were taken on the edges of the nanowires, at the thinner branches ($\rightarrow$ = 20 nm) to ensure transmission of the electron beam through the samples. Diffraction patterns were taken using beam widths smaller than the width of the nanowires, to minimize probability of outside contributions to measured patterns.

2.4. Nanowire conductivity

Electrical conductivity measures were taken on individual nanowires trapped across litho-graphically patterned electrical contacts [26]. In this process, a silicon wafer is masked with photoresist that has been patterned into an array of source-sink contact strip pairs. A source-sink pair consist of two, macro-scale, square contacts, each with a single lead approximately 4 mm in thickness and 10 mm in length extending towards the complimentary pad. The two parallel whiskers are separated by a 2 mm gap of patterned photoresist.

To prepare a single sample for testing, a suspension of nanowires in methanol is pipetted onto the substrate surface and the solvent is allowed to evaporate. Nanowires stick preferentially to the photoresist mask and not to the silicon wafer surface. Once dried, each source-sink pair is reviewed via SEM imaging to identify if any source-sink pair has a single nanowire trapped on the photoresist and also is bridged between the source lead whiskers and the sink lead whisker, Figure 5. Successful preparations are labeled, and the whole wafer is then sputter coated with Ti ($h = 5$ nm) adhesion and Au ($h = 50$ nm). After metalization, photoresist is lifted off, leaving behind the source-sink leads/contacts in gold, with bridging nanowires pinned between the Ti/Au layer and the silicon wafer.

Once prepared a voltage bias was applied across the contacts and current was measured through the leads. Electron transport measurements were performed using an Agilent 4156B semiconductor parameter analyzer. Current-voltage curves were generated to characterize nanowire resistivity.

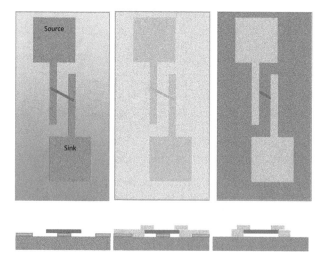

Figure 5. Schematics showing three steps used to MEMS fabricate nanowire resisitivity test system. A photoresist mask (blue) is spin coated and patterned onto a silicon substrate (gray) and nanowires are dispersed from suspension until a source/sink pair trap a wire. Metallization (center) with a gold layer followed by liftoff (right) leaves patterned leads holding the nanowires in place for testing.

2.5. Helium leak testing

Helium leak tests are a standard method for assessing electronic package hermeticity. Here we use helium leak tests to provide a preliminary assessment on whether our nanowire-template assemblies have hermetic properties worthy of microelectronics packaging applications.

In practice, helium leak tests are performed by subjecting two-dimensional test samples to vacuum on one side, while helium gas is introduced on the opposite side via gas gun. The vacuum side uses a mass spectrometer to measure any helium that has penetrated through the substrate/interconnects as a result of the applied vacuum. Helium concentrations detected can be correlated to leak rates.

Helium leak tests were performed using an ASM 182-TD (Alcatel, Inc.) helium leak detector with capability of detecting helium leak rates down to 5×10^{-12} mbar L s^{-1}. Typically, values below 1×10^{-11} mbar L s^{-1} are considered hermetic for microelectronics applications [4, 27, 28]. A custom, sample mounting fixture was designed for testing the nanowire embedded membranes. The fixture has an ultra-fine polished surface to ensure a proper seal between the test membrane and the fixture. The fixture was mounted to the inlet aperture of the leak detector using a standard vacuum seal and clamp. Samples are placed over top of a small circular inlet in the center of this fixture and held in place by the applied vacuum, a small amount of vacuum oil is applied to a polymer O-ring which is placed between the surface of the fixture and the sample to ensure a good seal.

To take measurements samples are mounted and vacuum applied until the flow rate settles below 1×10^{-11} mbar L s^{-1}. A continuous small dose of helium was sprayed from a gas spray gun on the top side of the sample exposed to ambient air. Helium gas was sprayed at the sample surface at a distance of 10cm with a pressure of 20 lbs for 10 seconds. After helium exposure, the highest observed leak rates were measured and recorded. Leak rate measurements were repeated three times for each sample. After each measurement, a time delay was given in order to return the leak rate to re-equilibrate.

3. Results and discussion

3.1. Platinum–Iridium nanowire electrodeposition

A series of nanowire-template samples were prepared using different plating solution pH: 1.8, 3.1 and 5.0. Subsequently, nanowires were isolated and examined via SEM to identify what pH produced preferred morphologies. For our application, high-aspect ratio and non-porous wires are desirable. Figure 6 shows a series of representative SEMs of isolated nanowires prepared from solutions of different pH, specifically (from top to bottom): pH = 1.8, 3.1 and 5.0.

Nanowires were first deposited at pH = 1.8, Figure 6a, isolated nanowires showed some mechanical compliance, evidenced by their visible bending in images. However, analysis of image series confirmed that at this pH, high aspect ratio nanowires could not be achieved. It is unclear what the root cause may be. Typically, at such low pH, hydrogen co-deposition played some role in impacting deposition. It is also possible that template dissolution under such high acidity may be taking place.

Figure 6. a) SEM micrograph of isolated nanowires with shorter lengths electrodeposited at pH=1.8, (b) SEM micrograph of isolated nanowires with brittle structures electrodeposited at pH=5 and (c) SEM micrograph of isolated dense nanowires with dense structures electrodeposited at pH= 3.1.

At pH = 5.0 Figure 6b, isolated nanowires were fragmented also with small aspect ratio. Porosity within the individual segments could not be confirmed, however, we suspect that either the nanowries themselves are brittle and fragment post-isolation, or that the deposition results in porosity which leads to fragmentation, once the structural support of the template is removed. It is unclear at this time what the mechanism for the resulting microstructure may be.

In contrast to both previous samples, high-aspect ratio nanowires were successfully synthesized using a plating solution with pH = 3.1, Figure 6c. Image analysis showed that aspect ratios from 20:1 to more than 50:1 were achievable using this approach. More work is needed to better understand what mechanisms are responsible for controlling the transport mechanisms taking place in the nano-channels. From these data, we determined pH = 3.1 would be capable of producing nanowires with preferred morphological structure.

3.2. Controlling Pt: Ir composition with potential

Co-deposition of binary alloys and multilayer nanowires has been reported elsewhere [29-31]. Here, platinum (Pt) and iridium (Ir) metal atoms are deposited through the AAO channels by electrochemically reducing from platinum ions $[PtCl_6]^{4-}$ and iridium ions $[IrCl_6]^{4+}$ in solution. By cycling the potential in a range below the equilibrium potentials for both ion complexes, the kinetics of deposition can be modulated to ensure that both elements are deposited in desirable quantities.

For this study, the potential range that produces a desirable Pt:Ir composition was unknown. We therefore selected five different potential ranges, Table 1, and prepared nanowires using each cycling range. All five potential ranges spanned 250 mV of potential but used different starting potentials (U_o) from $U_o = -0.5V$ to $U_o = 0.2V$ vs. Ag/AgCl.

Potential range (V) vs. Ag/AgCl	Average Pt:Ir ratio (%)
$\Delta U = 0.20$ to -0.15	68:32 %
$\Delta U = 0.02$ to -0.15	67:33 %
$\Delta U = 0.00$ to -0.15	62:38 %
$\Delta U = 0.00$ to -0.20	68:32 %
$\Delta U = -0.05$ to -0.15	85:15 %

Table 1. Deposition potential ranges and resulting PtIr nanowire fractional composition

Following deposition, nanowires were isolated and composition was tested using energy dispersive spectroscopy (EDS). Results in Table 1 show compositional fractions ranging from 62:38% to 85:15% platinum were attained by varying chemistry. While the exact mechanisms responsible for the differences in concentration are not well understood at this stage, we do know that shifting the potential range used for deposition affects the deposition kinetics inside the nano-channels.

It has been reported that platinum-iridium with 60:40 composition shows preferred properties for neurostimulation applications, e.g. highest electroactivity [32]. We therefore focused on further characterization of nanowires prepared with this deposition range ($\Delta U = 0.00$ to -0.15).

3.3. TEM characterization

TEM analysis was performed on 60:40% Pt:Ir nanowires to further characterize material properties. Figures 7a and 7b are bright-field and dark-field TEM images of platinum nanowires, respectively, deposited at the optimal conditions for creating hermetic AAO-feedthrough assemblies.

Figure 7a shows a gross image of a single isolated nanowire. Due to thickness and low magnification, no structural information (grain size, orientation, etc.) is distinguishable. At higher magnification and in dark field mode, however, grain boundaries and morphology can be distinguished. In Figure 7b, we estimate average grain size in the range of 5-10 nm, and they show no preferential growth orientations. There is some contrast observed between grains along the perimeter of the wire versus grains occurring within the central axis of the nanowries. This may suggest that contact with the AAO nanochannel surfaces may direct grain growth in some preferred orientations, however more studies are needed.

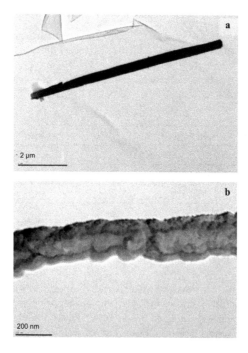

Figure 7. Bright-field (a) and dark-field (b) images of platinum nanowires.

TEM diffraction patterns of the deposited nanowires, Figure 8, showed concentric ring patterns, confirming the dark field observations that the nanowires were deposited with a polycrystalline structure, with no preferred orientation, and with an average grain size of 3-5 nm. Grain sizes calculated from x-ray diffraction patterns using Scherrer formula may underestimate grain size, as strain effects can impact patterns. The radius(r) of the diffraction rings varies with h, k, l as shown in equation (1):

$$r \; \alpha \; \sqrt{h^2 + k^2 + l^2} \tag{1}$$

where h, k and l are the Miller indices that represent the crystallographic plane. The results of the calculations using equation (1) showed that the rings correspond to the planes (111), (200), (220) and (311) which represent a typical face-centered cubic (FCC) structure (Figure 10).

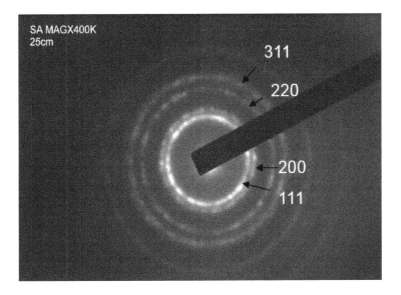

Figure 8. Electron beam diffraction pattern of platinum-iridium nanowire with fcc structure.

A HRTEM image of an individual platinum-iridium nanowire is shown in Figure 9. Since the values of the lattice parameters of platinum and iridium are too close to each other, the (−1−1 1), (1 1 1) and (0 0 2) planes labeled in Figure 11 represent the FCC crystal structure that may belong to either platinum or iridium (JCPDS 04-0802) or an alloy of the two. Consequently, the twin at the grain boundary might be due to the effect of a platinum or iridium alloying grain, indicating a bimetallic particle [33].

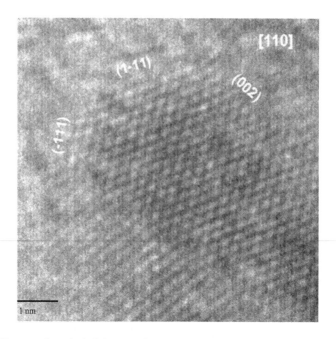

Figure 9. HRTEM image of an individual platinum-iridium nanowire

3.4. Conductivity measurements

Single nanowire, conductivity measurements were also performed on 60:40 Pt-Ir nanowires and compared against platinum nanowires prepared by a similar method. Figure 10a and 10b show scanning electron micrographs of lithographically fixated platinum and platinum-iridium nanowires, respectively, fixed between two lithographically patterned Au contacts. Platinum nanowires were used as a comparator as this is the only other known method of synthesizing similar nanoscale feed-through constructs. The method for their synthesis is described elsewhere [24]. Resistivity measurements were taken across the nanowire bridge and lead resistivity was subtracted out based on the bridging nanowire's location on the source and sink contact strip (approximately 2 μm for both samples).

Three representative current-voltage plots for both nanowire species are plotted in Figure 11. The slopes of these plots are inversely related to resistance [I = (1/R)V], therefore smaller slope magnitude suggests reduced resistance. For the six nanowires tested here we can calculated almost 2-fold improvement in conductivity for the Pt-Ir nanowires vs. pure platinum, which is consistent with known intrinsic properties for both metals (ρ_{Pt} = 105 nΩ m and ρ_{Ir} = 47.1 nΩ m, respectively). More investigations are needed in this space.

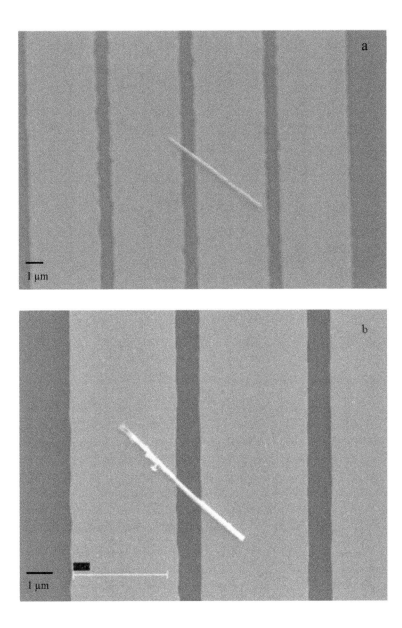

Figure 10. SEM micrograph of the testing device used for electrical conductivity measurement of a) single platinum nanowire and b) single platinum-iridium nanowire.

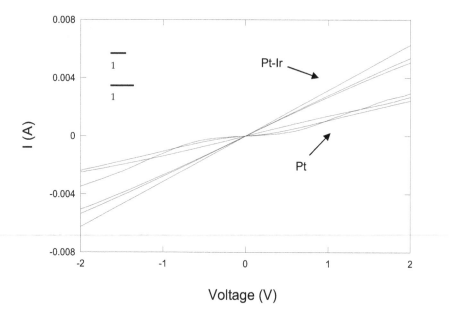

Figure 11. Current vs. voltage plots demonstrating the improved conductivity of platinum-iridium nanowires.

3.5. Helium leak testing

Pt-Ir nanowire in template assemblies were prepared with 60:40 (Pt:Ir) composition, and subjected to helium leak testing. Figure 12 shows a cross-sectional SEM micrograph of a fractured AAO template following platinum-iridium nanowire deposition. The conductive metallic nanowires are clearly seen as white, high-aspect ratio elements surrounded by the nano-channeled AAO (darker surrounding material). The right side of the image shows the 20 nm aperture side of the template and some of the bifurcations and branching can be resolved. The left side of the image shows the 200 nm aperture side of the template. Here it can be seen that complete filling of the channels has not been accomplished in this sample.

Helium leak testing results for these samples were taken an averaged 1.5×10^{-11} mbar L sec^{-1} (variance not calculated). These results suggest that these structures would meet industry standards for hermetic leak rates [27, 28], which have been reported at values as high as 2×10^{-10} mbar L sec^{-1}. For comparison, samples prepared using other potential ranges were also tested and found to be one to two orders of magnitude more leaky. These results suggest that it may be possible to fabricate hermetic feed-throughs using these types of nano-scaled constructs, however care must be taken to ensure that the metal deposition procedure produces non-porous, high-aspect ratio conducting elements.

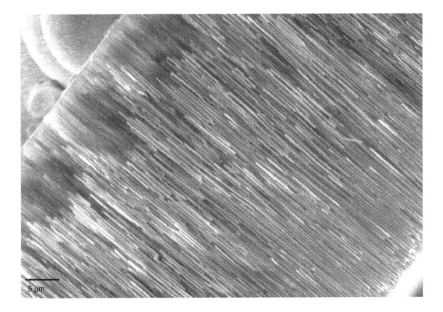

Figure 12. SEM micrograph of the cross section of platinum-iridium nanowires grown in AAO pores.

4. Conclusions

Platinum, iridium and alloys of the two will continue to play a significant role in biomedical devices due to their biocompatibility, resistance to corrosion, and electrochemical properties under physiological conditions. Processing challenges for these metals will continue to drive research and development to discover novel and efficient ways to create structures and components that meet technological demands. Namely, simpler and more reliable ways to fabricate components will always drive innovation in this space.

Hermetical packaging for implantable microelectronics will continue to use platinum and iridium feed-throughs embedded in non-reactive, electrically insulating templates/substrates. As the size of microelectronic devices decreases, and the number of feed-throughs per unit area of package increases, novel approaches to fabricating feedthrough technologies in an effective and economical way will be required.

Here we report on platinum-iridium alloy nanowires electrochemically deposited in nanoporous aluminum oxide using a template synthesis technique. We believe that series of nanowires, connected in parallel, can be used substituted for single conducting elements of equivalent cross-sectional area. However a careful understanding of how solution chemistry, deposition potential, and other parameters affect nanowires composition and morphology as well as feed-through hermeticity.

Hermeticity test results, as well as conductivity measurements, suggest that platinum-iridium nanowires may prove a viable platform for developing novel feed-through technologies. More investigation is needed to better correlate material properties of the deposited nanowires with resulting performance results. Additional work is also needed to develop ways of integrating these assemblies into entire chip package designs. However these results are promising.

Acknowledgements

Financial support was provided by the National Science Foundation under Grant No. EEC-0310723. The authors thank Dr. Chongwu Zhou, Haitian Chen and Yao Maoqing (Department of Electrical Engineering, University of Southern California) for their assistance with fabrication and electrical properties characterization.

Author details

Artin Petrossians[1], John J. Whalen III[2], James D. Weiland[2,3] and Florian Mansfeld[1]

1 The Mork Family Department of Chemical Engineering and Materials Science, University of Southern California, Los Angeles, California, USA

2 Department of Ophthalmology, University of Southern California, Los Angeles, California, USA

3 Department of Biomedical Engineering, University of Southern California, Los Angeles, California, USA

References

[1] Horch K & Dhillon G. S. Eds, Neuroprosthetics- Theory and Practice. Series on Bio-engineering & Biomedical Engineering. River Edge, NJ: World Scientific Publishing; (2004).

[2] Thomas, R. W. "Moisture, Myths, and Microcircuits." IEEE Transactions on Parts, Hybrids, and Packaging. 12(3) (1976) 167-171.

[3] Osenbach, W. Water-Induced Corrosion of Materials Used for Semiconductor Passivation. J. Electrochem. Soc. 140 (12), (1993). 3667-3675.

[4] Tummala, R.R. Microelectronics Packaging Handbook. Springer. 1988.

[5]   Ramesham, R, & Ghaffarian, R. Electronic Components and Technology 2000: Conference Proceedings, May 21-24, 2000, Piscataway, NJ, USA, IEEE; (2000).

[6]   Roy, S, Ferrara, L. A, Fleischman, A. J, & Benzel, E. C. "Microelectromechanical systems and neurosurgery: a new era in a new millennium." Neurosurgery (2001).

[7]   Roy, S, & Fleischman, A. J. "Cytotoxicity Evaluation of Microsystems Materials Using Human Cells." Sensors and Materials (2003).

[8]   Ferrara, L. A, Fleischman, A. J, Togawa, D, Bauer, T.W., Benzel, E.C., Roy, S. "An in vivo biocompatibility assessment of MEMS materials for spinal fusion monitoring." Biomed Microdev. 5 (2003) 297-302.

[9]   Johnson, L.J., Perkins, F.K., Merritt, C., Skeath, P. Justus, B., Weiland, J., Humayun, M., Scribner, D. "Low Charge Density Stimulation Of Isolated Retina With Microchannel Glass Electrodes." 2002 ARVO Conference. Fort Lauderdale, FL. (2002).

[10] Martin, C. R. Nanomaterials: A Membrane-Based Synthetic Approach. Science (1994).

[11] Yu-zhang, K, Guo, D. Z, Mallet, J, Molinari, M, Loualiche, A, & Troyon, M. Materials, Devices, and Systems III 2006: conference proceedings. October 2006, Boston, Massachusetts, USA SPIE, (2006). 2-3.

[12] Rothkina, L, Lin, J. F, & Bird, J. P. Nonlinear current-voltage characteristics of Pt nanowires and nanowire transistors fabricated by electron-beam deposition." Applied Physics Letters (2003).

[13] Sasaki, M, Osada, M, Higashimoto, N, Yamamoto, T, Fukuoka, A, & Ichikawa, M. "Templating fabrication of platinum nanoparticles and nanowires using the confined mesoporous channels of FSM-16- Their structural characterization and catalytic performance in water gas shift reaction." J. Mol. Catal. A-Chem. 141 (1999) 223-240.

[14] Sasaki, M, Osada, M, Sugimoto, N, Inagaki, S, Fukushima, Y, Fukuoka, A, & Ichikawa, M. "Novel templating fabrication of nano-structured Pt clusters and wires in the ordered cylindrical mesopores of FSM-16 and their unique properties in catalysis and magnetism." Microporous Mesoporous Mater. (1998). 597-606.

[15] Fukuoka, A, Higashimoto, N, Sakamoto, Y, Inagaki, S, Fukushima, Y, & Ichikawa, M. "Preparation and catalysis of Pt and Rh nanowires and particles in FSM-16." Micropor. Mesopor. Mater. 48 (2001). 171-179.

[16] Husain, A, Hone, J, & Henk, W. Postma Ch., Huang X. M. H., Drake T., Barbic M., Scherer A., Roukes M. L. Nanowire-based very-high-frequency electromechanical resonator." Appl. Phys. Lett. (2003).

[17] Yang, C. M, Sheu, H, & Chao, S. K.J. "Templated synthesis and structural study of densely packed metal nanostructures in MCM-41 and MCM-48." Advanced Functional Materials (2002) 17.

[18] Hippe, C, Wark, M, Lork, E, & Schulz-ekolf, G. "Platinum-Filled Oxide Nanotubes." Microporous and Mesoporous Materials (1999).

[19] Fu, X, Wang, Y, Wu, N, Gui, L, & Tang, Y. "Preparation of colloidal solutions of thin platinum nanowires." J. Matter. Chem. (2003).

[20] Kimizuka, N. "Towards Self-Assembling Inorganic Molecular Wires." Advanced Materials. (2000).

[21] Gurlu, O, Adam, O. A. O, Zandvliet, H. J. W, & Poelsema, B. "Self-organized, one-dimensional Pt nanowires on Ge(001)." Appl. Phys. Lett. (2003).

[22] Luo, J, Zhang, L, & Zhu, J. "Novel synthesis of Pt6Si5 nanowires and Pt6SiSi nanowire heterojunctions by using polycrystalline pt nanowires as templates." Advanced Materials (2003)., 5.

[23] Llopis, J. F, & Colom, F. Platinum. In:, Encyclopedia of Electrochemistry of the Elements. Bard A.J. (Ed). Marcel Dekker, New York. (1976).169-219.

[24] Whalen III J.J. Weiland J.D., Humayun, M.S., Searson, P.C. "Microelectrode System for Neuro-Stimulation and Neuro-Sensing and Microchip Packaging." US Patent Application. US2007/0187238.

[25] Petrossians, A. Whalen III J.J., Weiland J.D., Mansfeld F. "Electrodeposition and Characterization of Thin-Film Platinum-Iridium Alloys for Biological Interfaces." J. Electrochem. Soc. (2011).

[26] Chen, P-C. Ishikawa, F.N., Chang, H-K., Ryu, K. Zhou, C. "A Nanoelectric Nose: A Hybrid Nanowire/Carbon Nanotube Sensor Array with Integrated Micromachines Hotplates for Sensitive Gas Discrimination." Nanotechnology. 20 (2009) 125503.

[27] Davy, J.G. "Calculations for Leak Rates of Hermetic Packages." IEEE Trans. Parts, Hybrids and Packaging. (1975).

[28] Jiang, G., Mishler, D., Davis, R., Mobley J.P., Schulman, J.H. " Zirconia to Ti-6Al-4V Braze Joint for Implantable Biomedical Device." J. Biomed Mater Res Part B: Appl Biomater. 72B (2005) 316-321

[29] Shao, M. W. Chen, R. C. Cammarata, P. C. Searson, and S. M. Prokes, "Deposition and characterization of $Fe_{0.55}Co_{0.45}$ nanowires, " J. Electrochem. Soc. 154, D572-D576 (2007).

[30] T. S. Eagleton, J. Mallet, X. Cheng, J. Wang, C. L. Chien, and P. C. Searson, "Electrodeposition of Co_xPt_{1-x} Thin Films, " J. Electrochem. Soc. 152, C27-C33 (2005).

[31] M. Chen, P. C. Searson and C. L. Chien, "Micromagnetic Behavior of Electrodeposited Ni/Cu Multilayer Nanowires, " J. Appl. Phys. 93, (2003) 8253-8255.

[32] Holt-Hindle, P., Yi, Q., Wu, G., Koczkur, K., Chen, A. "Electrocatalytic Activity of Nanoporous Pt–Ir Materials toward Methanol Oxidation and Oxygen Reduction." J. Electrochem. Soc. 155 (2008) K5.

[33] Nitani, H., Yuya, M., Ono, T., Nakagawa, T., Seino, S., Okitsu, K., Mizukoshi, Y., Emura, Y., Yamamoto, T.A."Sonochemically synthesized core-shell structured Au-Pd nanoparticles supported on gamma-Fe2O3 particles." J. Nanoparticle Res. 8(6) (2006) 951-958.

Permissions

The contributors of this book come from diverse backgrounds, making this book a truly international effort. This book will bring forth new frontiers with its revolutionizing research information and detailed analysis of the nascent developments around the world.

We would like to thank Kenichi Takahata, for lending his expertise to make the book truly unique. He has played a crucial role in the development of this book. Without his invaluable contribution this book wouldn't have been possible. He has made vital efforts to compile up to date information on the varied aspects of this subject to make this book a valuable addition to the collection of many professionals and students.

This book was conceptualized with the vision of imparting up-to-date information and advanced data in this field. To ensure the same, a matchless editorial board was set up. Every individual on the board went through rigorous rounds of assessment to prove their worth. After which they invested a large part of their time researching and compiling the most relevant data for our readers. Conferences and sessions were held from time to time between the editorial board and the contributing authors to present the data in the most comprehensible form. The editorial team has worked tirelessly to provide valuable and valid information to help people across the globe.

Every chapter published in this book has been scrutinized by our experts. Their significance has been extensively debated. The topics covered herein carry significant findings which will fuel the growth of the discipline. They may even be implemented as practical applications or may be referred to as a beginning point for another development. Chapters in this book were first published by InTech; hereby published with permission under the Creative Commons Attribution License or equivalent.

The editorial board has been involved in producing this book since its inception. They have spent rigorous hours researching and exploring the diverse topics which have resulted in the successful publishing of this book. They have passed on their knowledge of decades through this book. To expedite this challenging task, the publisher supported the team at every step. A small team of assistant editors was also appointed to further simplify the editing procedure and attain best results for the readers.

Our editorial team has been hand-picked from every corner of the world. Their multi-ethnicity adds dynamic inputs to the discussions which result in innovative outcomes. These outcomes are then further discussed with the researchers and contributors who give their valuable feedback and opinion regarding the same. The feedback is then collaborated with the researches and they are edited in a comprehensive manner to aid the understanding of the subject.

Apart from the editorial board, the designing team has also invested a significant amount of their time in understanding the subject and creating the most relevant covers. They scrutinized every image to scout for the most suitable representation of the subject and create an appropriate cover for the book.

The publishing team has been involved in this book since its early stages. They were actively engaged in every process, be it collecting the data, connecting with the contributors or procuring relevant information. The team has been an ardent support to the editorial, designing and production team. Their endless efforts to recruit the best for this project, has resulted in the accomplishment of this book. They are a veteran in the field of academics and their pool of knowledge is as vast as their experience in printing. Their expertise and guidance has proved useful at every step. Their uncompromising quality standards have made this book an exceptional effort. Their encouragement from time to time has been an inspiration for everyone.

The publisher and the editorial board hope that this book will prove to be a valuable piece of knowledge for researchers, students, practitioners and scholars across the globe.

List of Contributors

Norihisa Miki
Department of Mechanical Engineering, Keio University, Hiyoshi, Kohoku-ku, Yokohama, Kanagawa, Japan

Fang Gang Tseng
Institute of NanoEngineering and Microsystems (NEMS), National Tsing Hua University, Hsinchu, Taiwan
Department of Engineering and System Science (ESS), National Tsing Hua University, Hsinchu, Taiwan
Division of Mechanics, Research Center for Applied Sciences, Academia Sinica, Taipei, Taiwan

Pen-Cheng Wang
Department of Engineering and System Science (ESS), National Tsing Hua University, Hsinchu, Taiwan

Tuhin Subhra Santra
Institute of NanoEngineering and Microsystems (NEMS), National Tsing Hua University, Hsinchu, Taiwan

Damien C. Rodger, James D. Weiland and Mark S. Humayun
University of Southern California, Los Angeles, CA, USA

Wen Li
Michigan State University, East Lansing, MI, USA

Yu-Chong Tai
California Institute of Technology, Pasadena, CA, USA

Ivanka Stanimirović and Zdravko Stanimirović
IRITEL A.D., Belgrade, Republic of Serbia

John Ojur Dennis, Farooq Ahmad and M. Haris Khir
Department of Fundamental and Applied Sciences, Department of Electrical and Electronic Engineering, Universiti Teknologi PETRONAS, Perak, Malaysia

Hamed Sadeghian
Technical Sciences, Netherlands Organization for Applied Scientific Research, TNO, CK, Delft, The Netherlands
Structural Optimizations and Mechanics Group, Department of Precision and Microsystems Engineering, Delft University of Technology, CD, Delft, The Netherlands

Fred van Keulen and Hans Goosen
Structural Optimizations and Mechanics Group, Department of Precision and Microsystems Engineering, Delft University of Technology, CD, Delft, The Netherlands

Kwun Nam Hui
Department of Materials Science and Engineering, Pusan National University, Geumjeong-Gu, Busan, Republic of Korea

Kwan San Hui
Department of Systems Engineering & Engineering Management, City University of Hong Kong, Kowloon Tong, Hong Kong, China
Department of Mechanical Engineering, Hanyang University, Seongdong-gu, Seoul, Republic of Korea

Gunasekaran Venugopal
Karunya University, Department of Nanosciences and Technology, Tamil Nadu, India

Sang-Jae Kim
Jeju National University, Department of Mechatronics Engineering, Jeju, South Korea

Artin Petrossians and Florian Mansfeld
The Mork Family Department of Chemical Engineering and Materials Science, University of Southern California, Los Angeles, California, USA

John J. Whalen III
Department of Ophthalmology, University of Southern California, Los Angeles, California, USA

James D. Weiland
Department of Ophthalmology, University of Southern California, Los Angeles, California, USA
Department of Biomedical Engineering, University of Southern California, Los Angeles, California, USA

Printed in the USA
CPSIA information can be obtained
at www.ICGtesting.com
JSHW011420221024
72173JS00004B/610